W9-BZB-347

The Mold Survival Guide

The MOLD SURVIVAL GUIDE FOR YOUR HOME

AND FOR YOUR HEALTH

JEFFREY C. MAY
CONNIE L. MAY

With a contribution by
John J. Ouellette, M.D., and Charles E. Reed, M.D.

The Johns Hopkins University Press
Baltimore & London

Note to the reader: This book is not intended to provide medical or legal advice. The services of a competent professional should be obtained whenever medical, legal, or other specific advice is needed. The authors make no warranty, either express or implied, regarding the recommendations offered or the practices described; nor do the authors assume liability for any consequences arising from the use of the content of this book.

© 2004 Jeffrey C. May and Connie L. May
Chapter Three © 2004 The Johns Hopkins University Press
All rights reserved. Published 2004
Printed in the United States of America on acid-free paper

9 8 7 6 5 4 3 2 1

The Johns Hopkins University Press
2715 North Charles Street
Baltimore, Maryland 21218-4363
www.press.jhu.edu

Library of Congress Cataloging-in-Publication Data
May, Jeffrey C.
    The mold survival guide : for your home and for your health / Jeffrey C. May and Connie L. May ; with a contribution by John J. Ouellette and Charles E. Reed
        p. cm.
Includes index.
ISBN 0-8018-7937-X (hardcover : alk. paper)—ISBN 0-8018-7938-8 (pbk : alk. paper)
    1. Molds (Fungi)—Control. 2. Molds (Fungi)—Health aspects. 3. Indoor air pollution. 4. Dampness in buildings. 5. Dwellings—Maintenance and repair.
I. May, Connie L. II. Title.
    TH9035.M39 2004
    616.9'6901—dc22   2003021414

A catalog record for this book is available from the British Library.

All photographs and photomicrographs were taken by Jeffrey C. May, except the scanning electron micrographs, which were obtained with his assistance at Severn Trent Laboratories, and the photographs on pages 34 and 35, which were provided by John J. Oullette, M.D. The photo on page 167 was kindly provided by Steve Goselin.

CONTENTS

PREFACE

You are reading this book, so you are probably worried about mold. Perhaps your basement has a musty odor or there are fuzzy dark patches on a bedroom wall, or perhaps you or someone in your family has mold allergies.

What is mold, and what can you do to prevent mold from growing in your home? This book will answer these and many other questions you have about mold.

And speaking of questions, who are we, to be writing about mold?

Jeff May

First, I'm a home inspector. Years ago, before people purchased a house, they looked around, and if they liked what they saw, they plunked down a deposit. They took their chances with the condition of the heating system, foundation, and roof. Today, buyers are more sophisticated, and most buyers hire a professional to do a home inspection before making a final commitment to a new home. A pre-purchase home inspection that follows the Standards of Practice of the American Society of Home Inspectors (ASHI) is very thorough, systematic, and informative, providing details about the home's construction, condition, and maintenance. In the fifteen years I have been in the profession, I have come to know thousands of homes, many more intimately than I cared to, and I have seen again and again how minor neglect can lead to decay and the need for costly repairs.

I wasn't always a home inspector; I was educated as an organic chemist (B.A. Columbia, M.A. Harvard). In addition, I am a certified indoor air quality professional (CIAQP); I have taken microbiology courses and trained other indoor air quality professionals. In my air quality work I combine a commitment to science and health with an interest in teaching and in solving building mysteries, such as mold and odor problems.

In the 1970s, when I was a high school physics and chemistry teacher, Sam Bass Warner, the parent of one of my students, gave me a microscope in a handsome case made of light-colored wood with a beautiful grain. The case had a comfortable black handle at the top and a hinged door with a jewelry-box lock at the front. Tied to a string on the handle was a small key. The microscope was nestled inside the box, held in place by curved brackets covered with green velvet.

I must admit I liked the case a lot, but the microscope didn't seem that special at the time, because I was more interested in molecules than cells. But I was also allergic to house dust, and I had two little children with asthma. One day it occurred to me that the microscope might help me discover if something in the dust was causing my family's problems. My first peek through the eyepiece revealed an invisible world, and I was hooked.

I now own three more microscopes, three digital cameras, and half a dozen air samplers. I can trap unseen particles from the air and view them on a computer monitor. I have personally investigated over 750 buildings and collected and analyzed over 15,000 air and dust samples. In the many hours I have spent bent over one microscope or another, I have learned what air and dust can contain: spores, pollen, pet dander, insect body parts and droppings, feather fragments, tire particles, soot, skin scales, cornstarch granules, cellulose fibers from clothing, wool dander from rugs, and pigment particles from spray-painting and laser printers, to name just a few.

In my lab I have microscope slides full of mold spores and petri dishes covered with mold growth. Despite my mold allergies, I am fascinated by the intricate beauty of growing mold. Fluffy blue or brown colonies fan outward like spreading drops of ink; black mold creeps beneath the petri dish's nutrient surface, dark bulges erupting here and there. Once I noticed a symmetrical colony of greenish mold turning dark as it was overrun by the explosive growth of another furry white mold. Under the microscope the "victim's" spores looked as if they had been smashed and cracked open by the aggressor.

Mold spores come in an amazing variety of shapes: some are round, some oval, and some spiral shaped. Some have smooth surfaces, and others have barbs. The number of spores produced by some mold colonies seems infinite, and the spores and the structures that bear them are arranged into a microscopic architecture with a never-ending web of order and complexity.

As a scientist I am interested in mold, but as an indoor air quality professional I am convinced that in over 80 percent of the houses I have investigated, elevated levels of mold spores in the air were responsible for some of the coughing, sneezing, runny noses, and breathing difficulties people were experiencing.

I am not a doctor and will not be giving medical advice; nor would I ever tell anyone to disregard medical advice in favor of environmental measures. If you are experiencing allergy or asthma symptoms in your home and you wonder if mold is the cause, see a physician. I would also discourage those who have mold allergies, or who live in households with people who have such sensitivities, to carelessly disturb what they believe might be mold growth.

Many people who react to mold suffer in silence not only their symptoms but also the unsympathetic attitudes and even disbelief of family members and friends. Understanding is key. Learning more about mold will help persuade disbelievers, convince sufferers that they are not imagining things, and turn everyone involved into more able warriors in the battle against this indoor contaminant.

Connie May

This is a book about mold, but it's also a book about scientific principles.

I always hated science. In high school I suffered through the required physics course, taught in a hands-on way that was supposed to be innovative and help people like me better understand the forces of the natural world. My classmates took careful notes while they watched the wave tank, and they got all excited about their magnets. But I just didn't get it. I cried while I pushed my brick around on its cart, and I flunked nearly every test.

In college I had to earn three credits in science, and they didn't let us substitute psychology, so I chose a biology course that started with one semester of botany and finished with a semester of zoology. In botany we alternated lab work and lectures with walks in the woods to collect plants, and I began to enjoy myself. Maybe science was OK after all. Then I hit zoology. My lab partner was a pre-med student with his own dissecting kit; I barely made it through the class alive.

It is one of the biggest ironies of my life that I married an organic chemist with a passion for science that doesn't stop at the laboratory door. During our courtship he talked about why shadows can contain aqua, yellow, and purple. He thinks about the adhesive properties of different kinds of paint. He puts petri dishes under spider webs to catch the arachnids' droppings so he can examine them under his microscope.

We met when I was teaching English and he was teaching chemistry in the same high school. Shortly after we got married he was hired to rewrite a chapter in a high school chemistry text, and he asked me to read some of his work. Guess what? I didn't get it. He tried to explain it to me. I still didn't get it. Finally he took the ideas apart and presented them to me step by step, starting at the beginning. At last I got it, and in the journey we took together, working on that chapter, we realized that science was not outside my mental grasp at all; rather, I thought differently than he did. I understood scientific concepts if they could be translated into my verbal, sequential way of thinking. At last the two of us could talk about why a rainbow has colors and why we feel hotter on a humid day. Now I not only appreciate the sights and smells and sounds of the world around me; I also understand what I am perceiving in a deeper way.

As we wrote this book together, I learned about mold, but I also learned about the movement of matter by exfiltration, infiltration, and diffusion, and the relationship of relative humidity and dew point to the phase changes of water. In most cases I can now recognize the different patterns of efflorescence, soot deposits, and mold growth. These subjects may not be at the top of your list of potential topics of dinner conversation (and I don't blame you a bit), but I feel more competent now, because I know when to worry about mold and when not to worry about it, and what I can do about it if I find it growing in my home.

In this instance, as in so many others, understanding scientific principles has given me tools to deal more effectively with some of the problems and worries of everyday life. I hope our book will help do the same thing for you.

Our Book

We are co-authors of this book, but Jeff is the building investigator and is thus the narrator; for this reason, when we say "I" in the chapters that follow, we mean Jeff. In part 1 of the book, "The World of Mold," we define what mold is (and is not) and discuss the conditions that contribute to its growth as well as how indoor mold growth can affect health. Chapter 3 in part 1, which describes how mold can affect human health, was written by Drs. John Ouellette and Charles Reed, experts in the health effects of mold exposure. In part 2, "The Search for Mold," we examine why mold grows in certain areas of a home and tell you how to find sources of contamination. The cleaning section, part 3, is last, because if you try to clean mold without understanding why it grows where it does, it will probably reappear and your efforts will have been in vain (and may even have been harmful).

ACKNOWLEDGMENTS

Many people have been supportive of our work, but there are a few to whom we'd like to give special thanks:

Michael Atwell, Steve Goselin, John Haines, Constance Jenkins, James Scott, and Davidge Warfield for offering invaluable feedback.

Cole Greenberg from Arizona, for his advice about evaporative coolers.

Larry Cerro from Florida, for his comments on air conditioning.

Family and friends who have listened to us talk about mold and mildew year after year, and who still agree to spend time with us.

PART I

THE WORLD OF MOLD

Mold growth may be part of the natural world, but that doesn't mean it's healthy to have mold flourishing on your wallpaper or in your carpet. In this first part of the book we define mold growth and discuss the effects that mold can have on health. We pay careful attention to providing scientific explanations, because you can't begin to control mold until you understand how it grows and spreads.

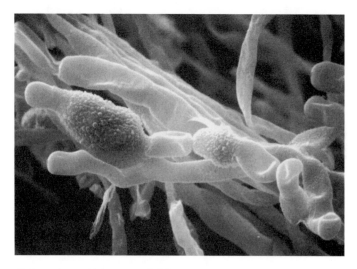

Cladosporium mold from a petri-dish culture. *Cladosporium* spores are often the most abundant type of spores found in air both indoors and outdoors. There are two fuzzy spores (magnified two thousand times) in this scanning electron micrograph (2,000x SEM).

Chapter 1

THE INDISPENSABLE KINGDOM OF FUNGI

One young couple called me because they were in a desperate situation. The husband, an attorney who had just come east from California for a new job, related their tale of woe. Two weeks earlier the couple had driven up at night to the house they were renting in a Boston suburb. Their moving truck was due to arrive the next morning at dawn, and they had expected the house to be clean and empty, ready to receive their belongings.

Unfortunately, the previous tenants had left the basement and attic completely full of rubbish. The attic insulation and basement contained animal waste, and black mold covered some of the basement walls. The couple knew that the carpeting, at least, had been washed, because it still felt damp; nonetheless, they were horrified. They knew no one in the neighborhood and didn't know where to turn for help so late at night. They called the police, and at midnight an off-duty officer volunteered his pickup truck and helped them remove the trash and take it to a nearby dump.

They spent that first night in the house, and when they woke up, the wife's arms were covered with flea bites. That was the last straw! They decided to find a new place to live, but meanwhile, the moving truck arrived. They had to unload their belongings, but they were worried about contamination. So they placed all their unopened boxes on sheets of heavy paper that they unrolled on top of the carpet. They never unpacked even one box.

They called me because they were breaking their lease, and they needed evidence to show there were problems in the house. I was as outraged by the situation as they were, and I took photographs and collected samples to document the conditions. But I was also curious to see if mold had grown in the paper that had been sandwiched between their boxes and the damp carpet below, so I looked at a sample under the microscope and was astonished to see that in just two weeks the fibers had become completely infiltrated by mold.

Fungi

In this book I will be using the word *mold,* which is a nonscientific term commonly used to describe just about any fungal growth. That said, molds, as well as yeast and mushrooms, belong to the kingdom *Fungi* (singular *fungus*)—one of the five kingdoms (or, in a newer classification ordering, two of the seven kingdoms) that also include the kingdoms of animals and plants.

Organisms in the animal kingdom take food into their bodies and use enzymes to digest it within. Fungi, on the other hand, do not have internal digestive systems; rather, they secrete digestive enzymes to break down living or dead matter in the environment. The products of this digestion are then absorbed and used for energy and for cell growth. While we don't want mold in our homes, fungi are an important part of nature's recycling system, reducing matter to simpler compounds that they and other organisms can use. Fungi clean the Earth by degrading dead matter, and in their search for nutrients they clear the forest floor of the fallen to make room for new life.

From Strength to Weakness

The expanding white cottonlike fuzz associated with mold growth is called the *mycelium* (plural *mycelia*), and it is made up of numerous rootlike threads called *hyphae* (singular *hypha*). The hyphae of liv-

Determining the relationships among living things is the monumental task of biologists. One of the principal ways of creating the needed organization is through the system of naming called *binomial nomenclature*. In this system, each organism is given a unique two-word Latin name. The first word, always capitalized, indicates the *genus* (plural *genera*). For example, human beings are in the genus *Homo*. The second word isn't capitalized and describes the *species*. When a particular species is named, both words are used. For example, all human beings are members of the species *Homo sapiens*, the only remaining living species in the genus.

So they can be adequately characterized, organisms are grouped into several other subdivisions. Related genera are arranged into *families*. Related families are gathered into *orders*, orders into *classes*, and classes into *phyla*. Each *kingdom* consists of a number of phyla.

Classification is a science that is in constant flux, and what was once the kingdom Fungi has now been reclassified into two kingdoms. In one, kingdom Eumycota, there are over a hundred species in the genus *Penicillium*. These species include *Penicillium roquefortii* and *P. camembertii*, used to make their respective cheeses, as well as *P. chrysogenum*, used to make the drug penicillin. (Note that when a species is named, the genus is often abbreviated to its initial—in this case, *P.*)

ing fungi are made up of elongated, water-filled cells. Digestion takes place mainly at the growing tips of the hyphae, so the hyphae have to keep branching and increasing in number in order to sustain the organism. As the hyphae digest, the material being consumed is transformed; that's how the fallen trunk of a once mighty tree is reduced to dust.

Many fungi grow on wood, whether alive (trees) or dead (the framing of a wooden building). Wood consists primarily of cellulose

Halloween horror. This Halloween decoration, left outside too long, was discovered during a home inspection. The head of the figure is an orange, with what is probably the mold *Penicillium digitatum* providing an eye and beard. The moldy orange sits atop a pumpkin that will soon be fodder for other microorganisms.

(a food source for many fungi) and cellulose-like compounds in a matrix of an amorphous dark polymer called *lignin.* These components of wood are what give trees the strength to grow to great heights and to bend but not snap in a strong wind.

If you view a sliver of wood through a low-power microscope, you can see that it looks like a bundle of straws with the straws segmented by porous cell walls or "caps." Each cylindrical shape between a pair of caps is a wood cell. Mold hyphae grow through the tube in the wood structure to break down the cellulose with an enzyme called *cellulase* and/or to digest the lignin with another enzyme called *ligninase.* The wood structure loses its strength as it is degraded.

Beneath the bark, in the outer part of a tree, called the *sapwood,* the living cells transport moisture and nutrients between the roots and the leaves. In the inner part of the tree, called the *heartwood,* the cells are no longer living; they contain air as well as oils, resins, tannins, and other wood compounds deposited there over time for protection from microbial growth. Though certainly not immune, heartwood is more resistant to microbial decay than sapwood, because the end-caps may be clogged, and the tannins, resins, and oils inhibit mold growth. Even so, most lumber, if allowed to remain damp in the presence of spores from appropriate wood-decaying fungi, will be weakened by the digestion associated with fungal growth.

Some fungi rot only dead trees, while others attack the wood in living as well as dead trees. Fungi that attack wood can be grouped according to how they digest the wood and by the color of the decay. A mold called *blue stain,* which often produces a blue discoloration, digests primarily the starch granules found in tree cells and used for food storage. Blue stain will therefore affect the health of the living tree, but because blue stain does not digest cellulose, it does not rot wood, nor do its spores commonly cause allergy symptoms, so the presence of blue stain on wood used for construction is not necessarily a reason for concern.

The decay caused by *brown-rot* and *white-rot* fungi can destroy the strength of wood in trees and buildings. Brown-rot molds digest the white cellulose but leave the wood brown, because they do not have enzymes to digest the brown lignin. (Indoors, brown rots such as *Meruliporia incrassata* and *Serpula lacrimans* can cause serious structural decay, but the spores rarely cause indoor allergies.) White-rot fungi can digest all the components of wood, including the lignin. As the hyphae spread through a piece of lumber, they digest the lignin into nutrients, thus eliminating the wood's brown color.

The infamous brown-rot fungus *Serpula lacrimans,* known commonly as *dry rot,* causes enormous damage to wood-frame buildings, particularly to structural components (joists, sills, and studs) on or close to the foundation, where the wood on masonry is often cool or damp. Under the right conditions the mold can reduce the wood to friable (readily crumbled) segments that seem to disintegrate under pressure. As they grow, some of the hyphae of *Serpula lacrimans* collect into thick, ropelike strands (sometimes up to a half-inch in diameter) that can be used like a pipeline to conduct water and nutrients. Thus, wood in seemingly dry areas is moistened by the water supplied from a remote location—one reason why this decay is (mistakenly) called *dry rot*.

The kind of wood used in construction often determines what kind of fungus grows and how quickly it grows. For example, some woods, such as white oak, black walnut, cedar, and old-growth longleaf pine (southern yellow pine), contain compounds that help the wood resist decay by inhibiting the chemical reactions the fungal cell needs to grow and reproduce. The framing of most homes built today consists primarily of woods such as Douglas fir or ponderosa pine that may not contain enough concentrations of tannins, oils, or other components to resist fungal growth. Roof and wall sheathing are made of *oriented strand board* (OSB)—chipped aspen (poplar) that is sprayed with resin, then heated and compressed into rigid sheets (also used to fabricate floor joists). When damp, these building materials succumb easily to fungal decay. (See table 1.)

Table 1. Resistance of Heartwoods to Fungal Decay

Resistant	Moderately Resistant	Slightly Resistant or Nonresistant
Cedar	Douglas fir	Aspens
Black cherry	Western larch	Birches
White oak	Eastern white pine	Pines
Black walnut	Longleaf pine	Spruces

Source: Adapted from Robert A. Zabel and Jeffrey J. Morrell, *Wood Microbiology: Decay and Its Prevention* (San Diego, Calif.: Academic Press, 1992), p. 401.

The presence of minor but chronic moisture in an older home may not result in much decay, but such dampness in a newer home might lead to major structural damage and even, in extreme cases, to failure of the framing.

Macrofungi and Microfungi

In a forest you might find fleshy fungi, or *macrofungi,* including brown-rot and white-rot fungi, on logs and tree trunks. The fruiting bodies of macrofungi, often called *mushrooms* or *toadstools,* are actually projections of a network of hyphae that may have digested wood from many feet away. Sometimes mushrooms seem to pop out of the soil or wood overnight, but they are part of a bigger living structure—a hidden mycelium that can spread under the ground over many acres, digesting dead plant and animal material along the way.

> The mycelium of a fungus called honey mushroom, or *Armillaria mellea,* has been found to occupy areas as large as 1,500 acres and to weigh an estimated 1,500 tons! (*Guinness Book of Records* [New York: Bantam Books, 1995].)

On a forest floor, in addition to mushrooms you would probably also be able to see powdery fungi, called *microfungi,* fanning out along the surfaces of fallen leaves. Microfungi also grow in abundance, though less visibly, throughout moist wood debris and soil, and they also have mycelia that consist of hyphae. Microfungi growing on the surface of wood do not usually penetrate the wood deeply enough to weaken it, though a type of decay called *soft rot* is caused by species of *Chaetomium*—microfungi that need a lot of moisture and that primarily attack wood that is buried in soil (such as the buried bottom of a fence post).

You don't have to be concerned about structural damage to wood framing if microfungi are the only fungi growing in your home. On the other hand, microfungi that produce large numbers of spores can have a serious impact on human health. These types of microfungi include most of the problem genera typically found indoors, what people call mold or mildew: *Cladosporium, Penicillium, Aspergillus, Alternaria, Stachybotrys, Chaetomium,* and others (though technically, when biologists use the word *mildew,* they are referring to those microfungi that grow on plants and cause plant disease).

Some species of *Chaetomium* grow on the paper of chronically damp drywall, where they produce a furry black layer full of allergenic spores. *C. globosum* emits a chemical that has a strong, earthy, musty odor (see "Moldy Odors" in chapter 4); in high concentrations, the chemical can irritate the eyes and mucous membranes. *Chaetomium* species are often mistaken for *Stachybotrys chartarum,* the so-called toxic black mold that also attacks saturated paper (see chapter 4). In newspaper and magazine articles *S. chartarum* is described as sometimes appearing dark green, but when it grows in homes it is always black. The dark green (*olivaceous,* or the color of olives) moniker comes from the biological description of *S. chartarum* when it grows in a petri dish in a lab. (Many microfungi produce black-appearing colonies, so black mold is not necessarily *Stachybotrys* mold, which does not grow on or in the thin layer of dust on tile, metal, or glass surfaces.)

Mold Spores

Most fungi reproduce by releasing spores. In the mushroom types of macrofungi, spores are located on the underside of the mushroom (so the edible portion of a mushroom consists in part of spores). Microfungi produce spore-forming structures, called *conidiophores,* all across the surface of the growth rather than just at specific locations. Usually the mold has a colored surface because of the concentrations of these spores, even though the individual spores and the conidiophores are microscopic.

The gills of a portabello mushroom. The top of the portabello mushroom is the white fleshy cap; the gills that hang down underneath are completely covered with microscopic spores, which are released into the air. We eat the gills and spores in salads and gravies: a fungal delicacy!

Like the seed of a plant, a spore has everything it needs to start new life, except moisture. Once the spore lands in a moist environment in the right temperature range, the components are activated and the spore begins to grow (germinates). Chemical changes within the spore may take place in minutes, though the first hyphal extension, called a *germ tube,* may not appear for up to several hours. The new growth (the germinating spore) is called a *germling.* If the conditions are right, over the next few days the germling will develop into a healthy mold colony (see chapter 2). Otherwise the germling will use up its stored food and die—though I have seen a dead germling with a lone hypha and single new spore at the end: a last gasp!

The relationship between living plants and their disease-causing fungi is a battle for survival. When a spore lands on a leaf, the spore releases enzymes, in some cases within two minutes, that dissolve the leaf's top protective, waxy coating. The germ tube appears and

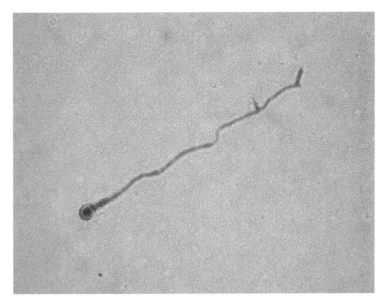

Cladosporium germling. The *Cladosporium* spore at the left germinated and lived briefly in the blower cabinet of a furnace. The single "daughter" spore produced is at the right, and the wormlike tube between the spores is the hypha. Under the right conditions this germling might have grown into a mold colony that would have produced millions of spores. (400x light)

secretes adhesives that firmly adhere the germling to the leaf. The leaf fights back by releasing toxic phenols as its "immune" response at the spore landing site, and, in turn, the spore's adhesives contain proteins that bind and deactivate tannins and phenols. Either the fungal growth is isolated or it spreads—and the leaf may die.

Most of the surfaces where spores will land indoors are "dead," so the battle is one-sided. As we'll see in chapter 2, all the spore requires are the proper moist conditions before it begins to digest our property.

The spores of microfungi are called *conidia* (singular *conidium*), and the structures that form them are called *conidiophores*. A fungal conidiophore may look like a microscopic version of a mushroom, and it serves a similar function: to disperse the conidia. Some conidiophores are brushlike structures such as those of the genus *Penicillium,* which produce conidia in long chains. The conidiophores of *Aspergillus* microfungi resemble balloons or lollipops and also produce spores in long chains. *Stachybotrys* spores are formed on the ends of a flowerlike structure, and the conidia grow stuck together by fungal mucilage in large clumps.

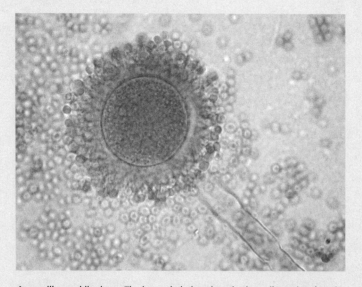

Aspergillus conidiophore. The large circle (a sphere in three dimensions) at the center and the diagonal stalk at the right are part of the lollipop-shaped conidiophore of an *Aspergillus* mold; in the background you can see many individual spores, which grow out in long, fragile chains like strings of beads. Because the spores are small, they are readily aerosolized. (About 1,000x light)

Chapter 2

WATERING MOLD AND SPREADING SPORES

Mold grows where it can find food: the dust on a bathroom ceiling, the starch paste on the back of wallpaper, or the plant fibers that make up the jute pad under a carpet. Just about anything organic can provide sustenance for mold growth. Mold has even been a problem in outer space. In the late 1980s the Russian space station *Mir* developed a musty odor. Species of the microfungi *Aspergillus, Cladosporium,* and *Penicillium* were found growing behind instrument panels.

What kind of mold grows in a particular place depends on what kind of food is available. Some species of *Aspergillus* digest leather, and species of *Chaetomium* and *Stachybotrys* digest paper. With the right food and temperature, moisture, and a little oxygen (from air), molds can thrive. Mold doesn't even need much air. Have you ever rolled up a wet plastic tablecloth or a damp tent and found it covered with mold when you went to use it again? The water cannot evaporate when the material is rolled up, and the little bit of air trapped in the folds is enough to allow microbial growth.

Mold spores indoors are inevitable, but if we keep dust levels low, repair leaks, and control indoor relative humidity, we minimize the conditions that can lead to fungal growth. Whenever mold is growing indoors, you know you have a moisture problem.

Moisture

Because moisture is essential for mold growth, it's important to understand the behavior of water in its various states. Air and water represent two states of matter: gas and liquid (solid is the third state). Many pure substances are familiar to us in any of these three states, or *phases*. For example, water can either be ice (solid), liquid, or gas (such as water vapor in air). Water in its liquid state can evaporate into vapor (a phase change), and from its vapor state it can condense back into liquid (the reverse phase change). Everyone knows that liquid water freezes into the solid state of ice, and ice melts back into liquid, but sometimes we ignore the obvious. For example, if you forget that ice and snowflakes are just the solid state of water, you may end up with a garage full of mold after parking a car inside that is caked with snow.

If we ignore leaks or unknowingly introduce water vapor into our homes, mold can grow on ceilings and walls, and even within ceiling and wall cavities.

Moldy boots. Mildew grew on these boots that were stored in a damp basement.

Leaks and Stains

Water can leak from a broken pipe or a plumbing fixture, through holes in a roof or cracks in a basement foundation, or around inadequate window or door flashing. The rate of leakage and how long the leakage lasts have a lot to do with how bad the mold problem will be.

Suppose water drips onto a floor for a few hours, and you wipe it up. Chances are there won't be a mold problem as a result. In a different scenario, let's say that a few drops of water an hour fall from a leaking pipe joint onto the upper surface of a drywall ceiling. Because the moisture is trapped in a relatively enclosed space, evaporation is slowed, and the drywall (gypsum sandwiched between two layers of paper) will remain damp. Mold may then grow, not only on the hidden, upper surface of the ceiling but also on the visible, lower surface, if the moisture soaks through.

When there are leaks like this, you will often see a circular stain on the ceiling before the mold starts to grow. The stain is made up of components in the building materials that dissolve in the water and are transported with the water to the lower surface of the ceiling. As the water spreads outward on the surface in the shape of a ring, it carries the materials with it. When the water evaporates, the stain remains behind. Green stains are often produced by copper in plumbing pipes. Brown stains can be produced when the water dissolves tannins in wood. The shapes of these stains provide a clue about where the water is coming from. Staining in and of itself is not an indication of mold growth, but there's always a good chance that where there is a stain, there is mold (growing on either surface of the drywall).

The growth of mold can also tell you a great deal about water flow. If you see a long patch of mold on a ceiling, water could be condensing along a pipe above and dripping down in a line onto the upper surface of the drywall. If you see a circular patch of mold, water is probably dripping from a pipe joint above. Bathtubs that overflow or toilets that leak at the seal can also create circular stains and mold growth in the ceilings beneath.

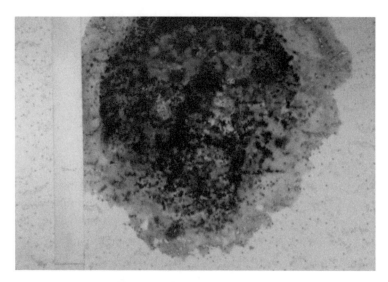

Mold growing on a ceiling tile. The tile is made up of minerals and paper (cellulose) fibers. The large black area in the center of the stain is mold. The lighter band at the perimeter is mold as well as colored material that could have come from water dripping through wood or moving through the fungal growth. A tile like this should not be disturbed without taking proper precautions (see chapter 10). The mold is probably digesting the cellulose.

Water moving through a porous material gives these stains their shape. A drop of water on a relatively homogeneous surface, such as a piece of cloth or paper, will spread out from the point of contact in all directions at the same speed, as a result of *capillary action*—a physical force that draws water into small gaps between fine solids, such as fibers or crystals. In materials with cracks and gaps, water will be drawn into these spaces by capillary action. Thus, on homogeneous drywall in a horizontal position (a ceiling), capillary water flow will create a circular pattern. If the ceiling has a crack (or seam), then the water will flow faster into the crack than through the drywall, and the staining will appear to originate from the crack. Gravity adds another force to leak patterns. If a leak occurs behind a vertical wall, capillary forces draw the water sideways into the paper and plaster, and gravity pulls the water down the wall. As the water soaks into porous wall materials, there is less liquid in the flow, and thus the pattern that results is an elongated stain shaped like a U or V.

Avoiding leaks prevents mold growth, and understanding how water flows can help you find the source of the leak.

Floor Floods

Stains from floor flooding create different patterns, although they are the result of the same physical forces: capillary action and gravity. If a hot-water heater breaks in a basement, for example, water will spread across the floor. Concrete is porous, so water will soak in by capillary action and drain out of any large cracks into the earth

The corner of a wet basement. The efflorescence (white mineral crystals) in the block foundation wall at the right and the black *Stachybotrys* mold growing on the far lower corner of the drywall at the left are signs of chronic moisture. The shape of the lighter surface pattern, which mimics the corner mold growth, outlines the pattern of capillary water flow originating from the corner.

Subtle signs of flooding. These are the diagonal stringers (tread supports) for base-ment stairs in a fairly new house. There was nothing stored in the basement and therefore no way to tell that there had been flooding, except for the two parallel ho-rizontal stains just above the first tread. A hallmark of floor flooding, such stains usually appear at the same height on each stringer.

below by gravity. If the water runs long enough, or if the rate of supply exceeds the rate of loss through the floor, the level of water will start to rise on the basement floor.

If you are wondering if there has been floor water in a basement, look carefully at the unpainted bottoms of the stair stringers and the exposed back of any drywall for stain patterns. A horizontal line at the same height on different surfaces often indicates that there has been water of significant depth in a basement; a series of lines will indicate more than one flood. On the other hand, a stain in the shape of an inverted U in wood or on drywall is an indication that water flowed up the surface by capillary action, and thus the flood may not have been extensive. (Keep in mind, though, that even pro-fessional home inspectors have been fooled by the appearance of supposedly "dry" basements that had been doctored by sellers who painted and cleaned these spaces prior to putting the properties on the market.)

Stains showing multiple events of basement floor water. The parallel stains and the mold on the drywall indicate multiple occurrences of basement floor water. The lines do *not* indicate the level of the water after each occurrence. Instead, they indicate the levels to which water rose by capillary action from the puddle that was on the floor.

Airborne Moisture

Moisture doesn't have to come from leaks or floods; it can also come from condensation of water on surfaces. I was once asked to determine the cause of severe mold growth on the window frames, walls, and ceilings of an apartment. Members of the family, all of whom had asthma, had bleached the mold, but it kept growing back. They thought that there was something wrong with the building, and they wanted the landlord to move them to a new apartment.

It was a winter day, and when I arrived at the apartment, steam was billowing from a large pot on the stove, damp clothing was hanging on a rack to dry in one of the bedrooms, all the windows were closed, and water was dripping down the inside of the windowpanes. The apartment was like a greenhouse for mold. In the center of the apartment, a closet containing the washing machine had a ceiling covered with brown mold colonies. Fuzzy yellow mold was growing on the bottom and sides of a lower bureau drawer. The plastered walls behind a bed and dresser, as well as the walls and ceiling of a closet, were black with *Cladosporium* microfungi, thriving in the fine layer of dust that typically collects indoors.

While this is an extreme example of conditions conducive to mold growth, it's not uncommon to find excessive moisture in an indoor environment. A shower adds about half a pint of moisture to the air for every five minutes of use. A clothes dryer, if vented into the house, adds about five pints of moisture per load. Cooking a meal for four people adds about a pint, and the breathing of four people produces just under half a pint per hour. And in some homes, people who feel they don't have adequate humidity add more water with furnace humidifiers (which can add up to a gallon of water per hour) and portable humidifiers.*

There are also building sources of moisture. In a newly constructed home, the concrete in the foundation walls and basement floor loses moisture for several months as it cures and dries. This can add up to ten pints of water per day to the indoor air. In a home built over a dirt crawl space, moisture evaporating from uncovered soil can add up to a hundred pints of moisture per day, some of which can find its way into the rooms above.

Whatever the sources, elevated levels of moisture in a home can lead to mold growth.

Relative Humidity and Dew Point

Liquid water evaporates into vapor, which is invisible. Indoor air that contains a lot of water vapor may feel muggy, yet it looks the same as dry air. (Similarly, sugar water tastes sweet and yet looks the same as plain water.) Steam in air is sometimes mistaken for water vapor, but the two are not the same. Visible steam consists of drop-

---

*W. Angell and W. Olson, *Moisture Sources Associated with Potential Damage in Cold Climate Housing* (St. Paul, Minn.: Cold Climate Housing Information Center, NR-FO-3405, Minnesota Extension Service, University of Minnesota, 1988).

lets of liquid water suspended in air. We can see these microscopic droplets because they reflect light. As the droplets evaporate into water vapor, the vapor mixes with the air and disappears from our view.

The *relative humidity* (RH) of air is a measure of how much water vapor is in the air at any given temperature, as compared with the total amount of vapor that could be in that air. For example, air that is at 50 percent RH has half the water vapor it could conceivably hold. If water evaporates from a surface into this air and the air temperature remains constant, the RH will rise.

As the temperature of air drops, the air's capacity to hold water vapor decreases. Indoor air in the winter at 70°F with 50 percent relative humidity can hold twice as much water vapor as it is currently holding. If the temperature of that air is lowered to 60°F, its capacity to hold water vapor decreases. Though the air is still holding the same *amount* of water vapor, the RH rises to 70 percent, so the air can now hold only 30 percent more water vapor. If the same air is reduced in temperature to 50°F, the RH would be 100 percent, and the air would be saturated. At this temperature, called the *dew point,* this air can hold no more water vapor. If more water vapor is introduced, vapor condenses (changes phase from vapor to liquid) on cooler surfaces (in this case, surfaces under 50°F).

The *dew point* is the temperature at which moisture will condense from air. Air at 70°F and 50 percent relative humidity has to be cooled to 50°F to reach its dew point, while air at 70°F and 20 percent RH has to be cooled to 25°F to reach its dew point. Air at 70°F and 70 percent RH reaches its dew point when cooled to 63°F. (A psychrometric chart provides the dew point for air at any given temperature and RH.)

Room air doesn't have to be at 100 percent relative humidity for water vapor to condense on surfaces, however. Air that comes into contact with a cold surface is itself cooled and may reach its dew point. Since the surface temperatures in a room are not uniform (some walls may be warmed by the sun, or others cooled because of heat loss due to missing insulation or the presence of a window), dew point conditions can exist near or on one surface of the room and not near or on others.

The patterns of mold growth can sometimes illustrate the history of temperature and moisture variations on a surface. Mold will grow on surface dust or wallpaper on cooler walls in rooms and closets as a result of high relative humidity, and on bathroom walls and ceilings as a result of condensation. Some molds require more moisture than others. A number of species of the microfungus *Aspergillus* can grow in 75 percent relative humidity and thus can thrive in humid conditions on dry-appearing surfaces. *Stachybotrys chartarum*, on the other hand, requires nearly saturated conditions, so it grows best on soaked paper (including the paper on drywall).

Accessible Moisture

When considering the potential for mold growth, you also have to take into account how much *accessible moisture* there is. Moisture accessibility is determined in part by the water content of the mold's food source. For example, a fresh apple, like most fruit, contains about 90 percent water and is thus readily subject to microbial decay, whereas dried apples are less subject to such decay and can be safely stored without refrigeration.

The level of accessible moisture is also determined by the *water activity*, or the way in which the water is physically and chemically bound to the substances in the food source. Mold will grow in a fresh plum, which contains approximately 90 percent water, yet mold will not grow readily in a dried prune made from that plum. The prune contains 28 percent water by weight—still a substantial amount of moisture that *should* support mold growth—yet prunes are not readily degraded by fungi, because the water is tightly bound to the carbohydrates (sugars, pectin, etc.) within.

Likewise, jelly contains about 35 percent water, but mold will not grow in jelly, because the high concentration of dissolved sugar reduces the water activity. Mold can grow on the *surface* of jelly, however, if the temperature of the metal cap is below the dew point of the moist air within the jar. This can happen because of temperature differences within the refrigerator itself, as the compressor cycles on and off, or because of warming of the jelly jar after it has been out on the table for a while and has then been put back into the refrigerator, where the cap cools faster than the jelly. Water then condenses on the inside of the cap and drips down onto the jelly, and the sugar becomes diluted at the surface. The water activity in that area increases, and some water becomes available for microbial growth.

Some dried foods can also support limited mold growth if water condenses on their surfaces. Not all foods with mold have to be thrown away. For a nonporous food like salami, removing a spot of mold and a small area around the spot may leave the rest edible.

Salt can also reduce water activity. For centuries meat (which is about 60 percent water to begin with) was preserved by "drying" it with salt. High levels of sugar and salt are natural deterrents of mold growth because they reduce the availability of water. Obviously you are not going to prevent mold from growing in your home by spreading salt and sugar on surfaces, but the concept of water activity can help us understand why mold will grow on some surfaces and not on others.

Airflows

One condominium owner, a semiretired teacher, spent many hours in his kitchen, preparing and eating meals and planning his classes. As the years passed he developed worsening respiratory problems and eventually required supplemental oxygen to breathe. I found that the refrigerator drip tray was overgrown with mold; whenever the refrigerator compressor turned on, air blew over the tray and carried spores into the room. In addition, the interior of the man's heat pump (supplying both heating and cooling) was full of growing mold.

This man's experience illustrates how air can carry unseen contaminants within a room or from room to room. For some reason, possibly because we sense odors so immediately, we understand and accept that an odor travels from its source to other rooms within moments, yet because we can't necessarily smell or see mold spores, it's hard to imagine that mold growth in a refrigerator drip tray can affect your breathing whether you are in the kitchen or in your bedroom, or that a mat of mold growing in the blower of a basement heat pump can produce allergens that can be spread by airflows in the duct system throughout the house.

Scientifically speaking, air and water are both fluids, and both flow. We can't usually feel air's movement, however, unless we stand in the wind outside or open a window for cross-ventilation. Still, air is in constant motion, even in the absence of perceptible wind. And air can carry chemicals (odors) and particulates (mold, pollen, bacteria, viruses) as it moves. When someone sneezes in an elevator, people understand that the cold virus may be transmitted by the suspended droplets dispersed into the air; likewise, when a mechanical system (heating, cooling, ventilation) "coughs" out microbes or allergens, people spending time in the space can be affected.

Infiltration and Exfiltration

Airflows can determine where moisture might condense. In a building, inside air leaks out and outside air leaks in through gaps around closed doors and windows. Inside air can flow into wall cavities through electrical outlets and openings around doors and windows, and into ceiling cavities around light fixtures. Outside air can flow into wall cavities through construction gaps. When air leaks from inside to outside, it's called *exfiltration;* the opposite airflow, from outside to inside, is called *infiltration.*

In a conditioned (heated and/or cooled) building, the temperature and relative humidity of the indoor air are usually different from these conditions outside. In the winter in a cold climate, the outdoor air is colder (sometimes by as much as 50°F or more) and drier than the indoor air. Warm air rises and cool air sinks, so the warm air from the house rises up and out of the building through exfiltration. At the same time, cold air leaks in (infiltrates) at the lower levels. In the summer in an air-conditioned building, the outdoor air is warmer and more humid than the indoor air, so colder indoor air sinks and exfiltrates at the lower levels, and warmer air infiltrates the house at the upper levels.

Let's imagine that we are sitting in a heated living room in Canada in January. Moisture from warm indoor air that exfiltrates (through electrical, plumbing, and other construction openings) into the wall cavities will condense on the cooler backside of the exterior wall and may foster mold growth. (In very cold climates, if enough vapor exfiltrates into the wall, ice can even form in the wall cavities. I've heard that water can start to leak out of the wall back into the room, soaking the drywall, if the weather warms suddenly and the ice within the wall cavity melts.) Insulation alone in wall cavities will slow but not stop this air leakage. The key to minimizing this problem is to keep indoor levels of relative humidity low (below 40 percent) and to stop moisture movement into wall cavities with the use of vapor barriers (such as sheets of aluminum laminate) or vapor retarders (thin sheets of plastic) on the back of the inside wall.

Now let's imagine sitting in an air-conditioned living room in Florida in May, where the humid outdoor air usually holds more moisture than the heated air in the Canadian living room usually holds—over five times as much. If this humid outdoor air infiltrates the wall cavities, moisture might condense on the cooler backside of the interior wall. In a hot, humid climate, a vapor barrier or retarder should be located where it will prevent outside vapor from entering the wall cavity; as long as it is airtight, the plywood sheathing normally used in exterior wall construction can serve as the vapor retarder.

Differences in Air Pressure

Air pressure differences can also help determine the pattern of airflows in and out of a house. Air pressure is a measure of the force exerted by molecules (the smallest particles in nearly all gases) when they collide with a surface. If you increase the number of air molecules in a given space, the air pressure will increase. When you inflate a balloon, you increase the number of molecules in the space, and therefore you increase the number of collisions per second (or the rate of collisions) between those molecules and the balloon's inner surface. As a consequence, the balloon expands. Temperature changes can affect air pressure. As the temperature of the air in a given space is *raised,* the air molecules move faster, and the air pressure increases within. If you *lower* the air temperature in a given space, the movement of the molecules slows, and the air pressure decreases (if you cool the balloon, it will shrink).

A house isn't a balloon, of course, but still, as the air within a house heats up, the air pressure increases, and the air will then expand and exfiltrate. As the air in the house cools, its pressure will decrease; the air will occupy less space, and outdoor air will infiltrate to take up the space.

Molecules in air or any other gas move from conditions of higher pressure to conditions of lower pressure. Returning to the living room in Florida, if we keep the inside air pressure slightly greater than the outside air pressure, exfiltration, rather than infiltration, will occur. If the inside air pressure is slightly lower than the outside air pressure, infiltration will occur, and humid exterior air may enter wall cavities. It is thus very important in humid climates to avoid having air pressure indoors that is lower than air pressure outdoors. This leads us to a discussion of how a hot-air heating or central air-conditioning system can create air pressure differences indoors.

A Balanced System

Differences in air pressure can create airflows not only from the outside to the inside (and vice versa) but also from one interior space to another, particularly in a home with a hot-air heating or central air-conditioning system, which consists of two types of ducts—supply and return—and a blower to move the air. Heated or cooled air comes out of the supply ducts, and room air is drawn into the system through the return ducts. The system is balanced if the same amount of air is being supplied as is being returned.

In the early years of hot-air heating systems, most rooms had a supply and a return. In most (but not all) newer construction, however, the rooms contain only supply ducts, with a central return or two located in a hallway. If the door to a room is closed, air is being blown in but not drawn out, thus creating increased air pressure in the enclosed space. (You can tell if the pressure in a room is too high when the blower is running if the door slams itself shut as you are closing it.)

Hot-air heat and air-conditioning supply and return ducts can be found anywhere in a house, but for the sake of this discussion, let's assume that an air-conditioning unit is in the attic of a one-story house. If more air is being supplied to the rooms than is being returned to the system, the air pressure in the habitable rooms will be greater than the air pressure in the attic and outside. In this case, exfiltration occurs in the rooms below the attic. If, on the other hand, more air is being returned to the system than is being supplied to the rooms, the air pressure will be lower in the habitable rooms than in the attic and outside, and air will infiltrate into the rooms.

Leaks in ductwork are the main cause of an unbalanced system. If a return duct in an attic is leaking, it will pull in some air from the attic and outside rather than from the rooms below. Exfiltration will then occur below, because the air pressure in the habitable rooms will be greater than the air pressure outside. If, however, the supply ducts in the attic are leaking, more air will be removed from the house than supplied, and infiltration will occur below the attic, because the air pressure indoors will be less than the air pressure outdoors.

No matter what forces are driving airflows, moisture carried by moving air can condense on cooler surfaces and foster mold growth. In a cold climate, an unbalanced system that is leaking warm, moist supply air out of heat ducts located in the attic can result in mildew growth in the winter, if moisture condenses on the cold attic sheathing. In the summer, cold air leaking from the air-conditioning system into the attic, though wasteful, may not result in mold growth on the sheathing, because the roof is too hot to allow condensation. On the other hand, because the air coming from the supply is usually below the dew point, moisture will probably condense on cooled surfaces near the air leak. This can lead to drip stains and mold growth on the attic floor as well as on the ceiling of the room below.

Old versus New

Exfiltration and infiltration aren't necessarily problems if the air is flowing around windows and doorways rather than through wall cavities. Exfiltration can allow pollutants and moisture to leak out of a house, and infiltration can introduce fresh air. In homes built before World War II, all the air exfiltrates and is replaced by exterior air in about an hour—an air exchange rate of 1. Newer homes are tighter and better insulated than older homes, so exfiltration and infiltration are reduced. A new home might have an air exchange rate of 0.25, meaning that a quarter of the house air is replaced by exterior air every hour. Thus, four hours would be required for a full exchange of the indoor air. While this may result in less condensation in wall cavities, it also means that normal activities such as showering and cooking can lead to high relative humidity levels in a new house. Under these conditions, condensation can occur in the home, and mold can grow on surfaces that are near the dew point. Likewise, when windows are replaced and insulation is added in older homes, moisture problems may appear for the first time.

Diffusion

Moisture can move *with* air by infiltration or exfiltration, and *through* air or even solids by *diffusion*—the random movement of any molecule through matter. The result of this movement is a more even distribution of the substance that is diffusing.

For example, if you place a wet sponge on a table, the water on the surface of the sponge will evaporate, and water from within will move to the surface by capillary action through the solid parts of the sponge and by diffusion through the pores. If you leave the sponge there for a day or two, it will dry out.

The human skin is always producing water vapor as moisture evaporates from the surface, whether we are sweating or not. If you wear porous clothing, the moisture can move through the spaces in the material by diffusion. If you wear a raincoat made of solid (non-porous) plastic (not nylon fibers), the moisture is trapped within because it cannot diffuse, and eventually it starts to condense. That's why you can feel clammy when wearing rubber or nonporous plastic outerwear.

Moisture diffuses through some materials more readily than through others. If you put a loaf of bread into a paper bag, the bread will dry out because the moisture in the bread will diffuse out through the paper. Bread sealed tightly in aluminum foil stays moist longer, however, because moisture cannot diffuse through metal (metal is a vapor barrier). People preserve leaves by pressing them between the pages of a book, where they dry out by diffusion. If a leaf were wrapped in plastic, it wouldn't dry out; in fact, it would probably get moldy, because plastic significantly retards vapor diffusion.

Why talk about diffusion of moisture in a book about mold? Because diffusion of moisture, particularly in hot, humid climates, can cause significant mold problems. If there is no vapor barrier or retarder in an exterior wall, air will infiltrate the wall cavity from the exterior, and water vapor can diffuse from the wall cavity into the house through the drywall. Because vinyl wallpaper contains a layer of solid plastic, it can act as a vapor retarder, minimizing the diffusion of moisture into the room, but the wallpaper also traps moisture in the drywall, often leading to mold growth. This has been an enormous and expensive problem in air-conditioned hotels located in hot, humid climates. Nonetheless, the amount of moisture moving into and out of buildings by diffusion is far less (one-tenth to one-hundredth less) than the amount that is carried by infiltration and exfiltration.

In the next chapter, Drs. John Ouellette and Charles Reed look at how mold affects health.

Chapter 3
HOW MOLD AFFECTS OUR HEALTH

John J. Ouellette, M.D., and Charles E. Reed, M.D.

Before the oil embargo of the 1970s, the need to conserve energy was not of great concern. But with that rude environmental awakening the housing industry began a major effort to reduce the amount of energy needed for heating and cooling. This effort included increased insulation and tighter construction with decreased ventilation, as well as new building materials and new products. Builders applied these new construction techniques not only to new homes and the remodeling of older homes but also to schools and office buildings.

An unintended important effect of these changes was to increase the heat and moisture content of the houses, often creating pockets of dampness where water-dependent microorganisms—molds and bacteria—thrived. In a sense, houses became incubators capable not only of generating airborne products of the growth of molds and bacteria but also of adding breakdown products of the rotting house. In some situations significant concentrations of volatile organic compounds released from building materials became evident. Lack of fresh air ventilation increased the concentrations of these contaminants in the air. This combination of events causing bad in-

31

door air quality often provoked illness in the people who lived in these houses. Properly built and maintained energy-efficient houses do not cause illness. However, the complexity of construction provides many possibilities for mistakes that do cause leaks or condensation.

Our involvement with this problem began after a Wisconsin Governor's Task Force reported the results of a study of TriState homes in 1986. In the early 1970s the TriState Company in northern Wisconsin began to manufacture prefabricated houses. These houses were affordable and energy efficient and so were very attractive to young couples with children starting out with their first homes. These prefabricated homes were widely scattered throughout the northern parts of Wisconsin, Michigan, and Minnesota and were spread out so far from one another that there was very little communication among the occupants. However, after several years it became obvious to these owners that there was something seriously wrong with the construction of their houses and that the occupants seemed to have more than the usual amount of respiratory illness. As a result of these complaints the governor of Wisconsin appointed a task force that included members of the University of Wisconsin Medical School, the Wisconsin Department of Health, and building scientists from the United States Forest Products Laboratory in Madison.

The task force found the main building flaw to be improper placement of the moisture barrier on the cold side of the wall cavity. This resulted in condensation of water in the walls, which promoted a lush growth of microorganisms that rotted through the walls. This impervious membrane on the outside of the walls also caused an influx into the interior of the house of the products of the decay. The air exchange rate was half as much as that in comparable conventional stick-built houses. Health questionnaires and medical examinations of the occupants of the TriState houses confirmed increased frequency and severity of respiratory illness, compared with neighbors living in other types of houses. Over the past few years there have been many other scientific studies in many countries linking unusual frequency of respiratory infections, allergic rhinitis and asthma, and other respiratory diseases to damp, moldy indoor environments.

People, Pollutants, Pathways, and Pressures

In understanding the complexity of the medical conditions arising from damp, moldy houses, we should look at forensic engineer Joe Lstiburek's four p's: *people, pollutants, pathways,* and *pressures*. It takes *people* to become sick; it takes *pollutants* to make them sick; the amount of exposure to the pollutant depends on the *pathways* that the pollutants take through the house, resulting from different *pressures* in different locations inside the house.

Concerning the *pollutants,* many different microorganisms grow in damp water-damaged spots; different organisms can cause different health problems. In any one location the particular variety of organisms may change as the amount of moisture, the temperature, or other conditions change. Mature molds produce spores for propagation. These spores can identify the particular species of mold, and spores are readily found in the air, allowing specific identification of the contaminating mold. It cannot be assumed, though, that the spores themselves are the only (or even the major) carriers of allergens or other molecules that cause disease. Many of these molecules are secreted into the water and surfaces where the organisms grow. When the area dries, these protein molecules become part of the dust. They can be detected in the air only by complex immunochemical tests. Bacteria grow where there is actual accumulation of puddles of water; molds grow where the spot is fairly constantly damp. Although most attention has been given to molds, there is increasing concern that although these bacteria do not cause infections, their products are important in causing respiratory diseases.

Pathways and *pressures* determine the amount of exposure. It is important to look at the house with a "whole house" approach, because one part of the house will affect another part of the house. For example, if there is a leak in a heating supply duct that goes through a crawl space, the pressure generated from the furnace blower will

The outer wall (cladding) of this house has been removed, exposing the plywood covering. The upper part of the structure, which corresponds to the attic of the house, is black with mold. The mold grew because the moisture generated in the house was unable to leave the wall because of the impermeable membrane placed on the cold side of the wall assembly. Moisture barriers should not be placed on the cold side of the wall assembly in cold climates. This flaw has been responsible for multiple problems in prefabricated or manufactured homes.

cause the air in the crawl space to be blown to other parts of the house. If there is mold growing in this space, then the mold will be distributed to other parts of the house as well.

Another example would be a north wind blowing on the north side of the house. Air would enter, or infiltrate, on that side of the house, possibly carrying the mold that formed inside the north wall into the occupied part of the house. The air would leave the house on the south side and possibly carry water in the form of water vapor into the south wall. The reverse would happen with a south wind: water vapor could be carried into the north wall, where it could condense on the cold wall and be available for future mold growth. Air is constantly moved through the house by a combination of pressures generated by the wind outside and the air handlers (blowers) inside, if they are present. If there is a pollutant generated in one part of the house, pressure differences can move pollutants throughout the house, thus making them available to the occupants in different rooms. Other sources of pressures that move air from one part of the house to another include forced-air heating and air-conditioning systems (as mentioned in chapter 2), as well as clothes dryers and exhaust fans in the bathroom or over the stove.

People are all different; everyone is unique. The type and severity of an illness that a person may develop depend on three independent variables: (1) the identity of the organism and the products it makes; (2) the amounts of the products in the air and the duration of

On the inner wall (wall cavity), the studs show water staining, and the insulation has decayed. There is some water damage on the plywood that occurred because the moisture barrier prevented the water vapor from passing through the outside wall. The water vapor condensed in the cold wall cavity, causing the water and mold damage.

exposures; and (3) the biological responses of the individual's cells to the products. Different people in the same house often respond differently, and any one individual may become sick from a combination of responses to several organisms acting through different biological pathways in the body.

Innate and Acquired Immunity

It is not at all surprising that most of the illness from water-damaged homes involves the nose and lungs. After all, they are the organs that are exposed to things in the air. The nose is an amazingly effective filter. It removes the bulk of airborne particulates and protects the lungs from all but the smallest particles. Only very tiny particles (called *respirable* particles by lung physicians) or volatile chemicals can pass on through the nose and the airways down to affect the lungs. But the nose pays a price. If the particulates carry allergens or irritants, the nose bears the brunt of the assault. And if the nose is plugged up and the person is breathing through the mouth, even large particles do reach the lungs.

Table 2 lists the specific diseases related to indoor bacteria and mold exposure in homes. The most frequent is upper respiratory inflammation, often progressing to chronic sinus infection. Symptoms are stuffy nose, mucus drainage, and discomfort in the sinus areas. In children especially, ear infections are frequent. Irritation and swelling of the mucous membrane in the nose make the individual especially susceptible to common colds and other respiratory viral infections. And the effects of indoor airborne irritants like cigarette smoke, nitric oxide from gas stoves, and volatile organic compounds are amplified.

Airborne particles of bacteria and molds cause this inflammation not by infection when they deposit on the membrane, but by a process known as *innate immunity*. Innate immunity is an immediate and nonspecific defense against invasion by viruses, bacteria, and molds. It developed millions of years ago in primitive invertebrate animals and is still the main defense for insects, shellfish, and similar creatures. It is an important early defense in humans. Cells in the respiratory mucus react directly with the invader and generate alarm substances that recruit a "SWAT team" of white blood cells to destroy the invader. Irritation, swelling, and mucus secretion result from this response. *Acquired immunity* to specific infections evolved more recently and is present only in vertebrates. It has developed further complexity and efficacy in mammals. Acquired immunity requires several days to produce antibodies and toxic lymphocyte blood cells specific to the invader. There are five classes of antibodies with overlapping functions. One of them, called *immunoglobulin E,* which is particularly effective against tropical parasites, causes allergy.

The close similarity in mice and humans of both innate and acquired immunity has made it possible to study the cellular and molecular details of both processes in mouse models, research that is not possible in humans. As a result of this research, immunologists now understand the biology quite well. Stimulation of innate immunity activates cells in the airways to secrete molecules called *chemo-*

Table 2. Diseases from Indoor Exposure to Bacteria and Molds Growing in Damp, Water-Damaged Homes

Chronic rhinitis and sinusitis, often with nosebleeds

Unusually frequent respiratory infections

Headaches, malaise, and/or low-grade fever

Allergic rhinitis and asthma

Hypersensitivity pneumonitis

Suspected but not yet scientifically established:

    Hemorrhagic pneumonia in infants

    Mycotoxin effects on brain function

    Mycotoxin effects on digestive tract function

    Endotoxin-induced chronic obstructive lung disease

kines, which attract blood cells to the site, and other molecules called *cytokines,* which activate the blood cells to accomplish their function: to destroy the invader. But in the circumstances we are discussing here, this innate response to bacterial and mold products does more harm than good. It causes swelling and mucus secretion with resulting discomfort from inflammation of the nose. In addition, it makes the person more susceptible to viral infections. Cytokines, particularly those called *tumor necrosis factor alpha,* generated in the respiratory tract during innate immune stimulation, circulate in the bloodstream to the brain and other organs, causing fever, headaches, and general malaise. They also activate the liver to produce a variety of acute-phase protective molecules.

Both bacteria and molds are capable of stimulating innate immunity. At present, bacteria appear to be the main airborne source of activation of the innate immune pathways and are especially likely to be growing in places where there is actual fluid water, not just dampness. Pools of condensed water in heating and air-conditioning ducts or the reservoir of some types of cold-water humidification systems frequently become contaminated by gram-negative bacteria, typically *Pseudomonas aeruginosa.*

Much more information is available from occupational exposures than from exposures in the home. Heavy acute exposure to airborne bacterial endotoxin (lipopolysaccharides) in farming and grain handling can cause *grain fever.* Chronic exposure to endotoxin in other industries that generate contaminated dust or mist is an important cause of the lung disease *emphysema.* The suspicion that emphysema may be related, at least in part, to contamination in homes has not been confirmed.

At least some of the bacterial and mold products that activate innate immunity have been identified and can be measured in air and settled dust. Common stimuli of innate immunity include *lipopolysaccharides* (endotoxin) and fragments of DNA from bacteria and to a lesser degree substances called *beta glucans* from mold hyphae. However, despite reasonably good scientific understanding of the process of inflammation from innate immunity, there are as yet no established blood or other tests to diagnose the condition. Diagnosis depends on the correlation between symptoms and exposure determined by careful examination of the home (see part 2).

Allergy

In susceptible individuals, mold products also stimulate allergy, a special form of acquired immunity characterized by production of a special type of antibody, *immunoglobulin E (IgE)*. Mold proteins react with IgE bound to cells in the nose and lungs. This stimulation leads to allergic rhinitis (hay fever) and asthma. How severe these diseases are depends on the amount of the exposure and the degree of the patient's response. However, chronic low-level exposure leads to more severe and persistent problems than occasional heavy exposure does. Molds make many different protein allergens. Some of these proteins are in the spores, others in the hyphae, but many of the most important are enzymes secreted into the environment as the mold grows. People with rhinitis and asthma are almost always allergic to several different allergenic molecules from the same organism and also allergens from several different organisms (like pollen and pets).

Other allergens in the home are important too, especially the tiny animals called *mites*. Mites grow in mattresses and upholstery, or in corners of the house where relative humidity is 70 to 80 percent and there is food (note that *Dermatophagoides farinae* grow in slightly drier places). Different kinds of mites eat different food. *D. farinae* means "wheat eating," and *D. pterynissinus* means "feather loving." Some species of mites eat skin scales, and other species forage on mold. Most molds require greater humidity than mites do, but some mite genera other than *Dermatophagoides* are often found wherever mold is growing.

Combined exposures to several allergens are likely to generate more severe disease. Diagnosis of indoor mold allergy relies to some extent on the demonstration of specific IgE antibody to the mold. At present, blood tests are more reliable than skin tests, but both are useful. Correlating the symptoms with the exposure is another important criterion for the diagnosis, and the diagnosis of asthma can be made more objective by adding serial breathing tests.

Of course, innate immunity and allergy are independent of each other. They are responses to very different stimuli, and the inflammation they cause in the airway involves a different combination of cells. But they can occur together, and when they do, they make each other worse.

Recently another form of allergic (or quasi-allergic) disease from mold has been recognized in some patients with nasal and sinus polyps and asthma. *Alternaria* species, growing on the surface of the mucous membrane of the nose but not an invasive infection, excrete a digestive enzyme that generates an allergic type of inflammation independent of IgE. This response has not been linked to indoor mold exposure, however.

Lung Infection and Disease

Some molds can cause lung infections such as *histoplasmosis* or *coccidiomycosis,* but these molds grow in the soil, not indoors. *Aspergillus fumigatus,* an indoor mold that grows well at body temperature, colonizes the lungs in some vulnerable patients with asthma or cystic fibrosis; it also causes generalized infection in patients with severe immune deficiency. These infections are not related to contaminated homes.

Hypersensitivity pneumonitis is a rare chronic lung disease that occurs most often as a result of occupational exposures. A characteristic form is farmer's lung from heat-loving bacteria known as *thermophilic actinomycetes* growing in moldy hay or silage. In homes the main causes of hypersensitivity pneumonitis are pet birds and neglected humidifiers contaminated with molds or actinomycetes. Several cases of hypersensitivity pneumonitis have been linked to molds or actinomycetes growing in water-damaged homes. This disease results from a different form of acquired immunity that depends on antigen-specific cytotoxic lymphocytes rather than on IgE or immunoglobulin G antibody. X-ray and breathing tests can usually confirm the diagnosis of hypersensitivity pneumonitis, but occasionally examination of the cells obtained by a bron-choalveolar lavage is necessary. Blood tests for specific antibody can usually confirm the causative organisms and can be related to the organisms identified in the home.

Some molds produce mycotoxins (see chapter 4). Exposure to the mold *Stachybotrys chartarum* and its mycotoxins, *trichothecenes*, has been linked to an outbreak of hemorrhagic pneumonia in infants living in Cleveland, although the conclusion that the myco-toxin actually caused the hemorrhagic pneumonia has not been scientifically proven. Several other mycotoxins cause disease in animals and occasionally in humans, if they eat foods heavily contaminated with molds. There have been case reports relating dementia, vomiting, and other symptoms to airborne exposures, but whether airborne mycotoxins pose a serious threat is still being debated. Further scientific investigation is needed to establish whether these mycotoxins are present in the air in sufficient concentrations to cause disease. Even if mycotoxins are present in damp, moldy homes, it is not certain that a sufficient amount of mycotoxin would be absorbed from the respiratory tract to circulate to other organs and cause a toxic effect. Assays for many mycotoxins are available, and quantitative animal challenge studies are feasible. So the question is likely to be answered in the future.

Chapter 4

MOLD WAGES BATTLE
Then and Now

A physician referred one of his patients to me, a woman who had inexplicably suffered several attacks of swelling and anaphylaxis while she was in one room of her house. She was diagnosed with angioedema, a sudden allergic reaction associated with severe swelling of the lips and face, among other symptoms. Because the illness is so unpredictable and the causes so mysterious, its complete name is *angioneurotic edema*. But in this case (as in probably so many others), there was nothing neurotic about this woman. Water had leaked into one wall, and mold had grown within. Wherever I tapped that wall for air samples, I found high levels of spores from *Aspergillus* mold. Once the *Aspergillus* mold sources in the room and in a few other areas of the home were eliminated, the woman's bouts of swelling and anaphylaxis stopped.

Mold isn't a recent scourge. According to one interpretation of the Bible, Leviticus warns people to demolish homes that are contaminated with mold. Farmers have known for centuries that mold can destroy crops: hay will rot from mold when it has been baled "green," or insufficiently dried out, as will grain if it goes into the silo with too high a moisture content. Ingesting mold-contaminated fodder can sicken animals. For example, during the 1930s the Russian economy suffered a great blow when large numbers of horses, crucial for military operations, died in the Ukraine from internal bleeding, sometimes within twenty-four hours after symptoms first appeared. These animals had ingested mold-infected hay. In 1960 in Great Britain, one hundred thousand turkeys died in a mysterious outbreak eventually called *turkey X disease*. Ultimately scientists linked the disease to the toxins produced by the microfungus *Aspergillus flavus*, present in the peanut meal fed to the turkeys.

Some researchers believe that the fungus *Claviceps purpurea*, or ergot, which grows on rye and is then eaten in the bread made from the grain, caused the delusions that led to the witch trials in Salem, Massachusetts, in 1692. Earlier, in the Middle Ages, a terrifying disease called *St. Anthony's fire*, also caused by ergot ingestion, struck whole towns. The disease's victims suffered fits, convulsions, and hallucinations, and their toes, ears, and fingers turned black. "Epidemics . . . have left bizarre accounts in which the screams of the dying, the stench of rotting flesh, and limbs actually falling off, are recorded in grisly detail."* Centuries later, in 1951, St. Anthony's fire broke out in a small French village, where more than a hundred people became ill with nausea, vomiting, and vertigo. Some suffered hallucinations and deep depression.

*Bryce Kendrick, *The Fifth Kingdom* (Waterloo, Ont.: Mycologue Publications, 1985), p. 289.

Whether allergic to mold or not, a person can become ill within twenty-four hours after inhaling large amounts of dust contaminated with microbial growth (such as spores and hyphae from *Fusarium*, *Penicillium*, or *Aspergillus* species). Called *organic dust toxic syndrome*, or *ODTS*, this illness is characterized by fever, coughing, and shortness of breath and is common among workers who clean out very moldy storage bins (such as grain silos) without using respiratory protection.

One of the better-known catastrophes caused by mold is the infamous potato famine in Ireland between 1845 and 1847, when the fungus *Phytophthora infestans* wiped out nearly the entire Irish potato crop. Hundreds of thousands of people starved to death, and millions more left the country to avoid starvation. In ten years the population of Ireland was halved, from eight million to four million.* Though Ireland was devastated, the influx of Irish immigrants to the United States had a lasting impact on the culture and politics of this country.

Mold's Arsenal

Molds have a number of weapons and devices they use to interact with other organisms and to affect human health.

Moldy Odors

Some people believe they can get a headache simply by being exposed to mold odors. The musty smells that some fungi emit may be unpleasant to humans, but they are useful to insects. For example, termites find moist, moldy wood by following the odor trail. Mold odors can also benefit the fungus. At the same time it produces spores, the fungus ergot growing on a rye plant secretes a discharge that contains sugars and that emits an odor similar to the smell of decaying fish. Flies, attracted by the odor, feed on the ooze. Spores stick to their bodies, and when they land on another healthy rye plant, they transfer the spores.

The chemicals in these odors are called *microbial volatile organic compounds* (volatile substances are those that evaporate at about room temperature), or *MVOCs*. You may not be familiar with the term MVOC, but you have certainly experienced these odors, as they are the chemicals that produce the characteristic smell of some molds.

Octenol, a common musty-smelling MVOC, attracts mosquitoes and other insects, which again may help to disperse the spores. One

*Kendrick, *The Fifth Kingdom*, p. 18.

Octenol (1-octen-3-ol) can be detected at its odor threshold (the lowest detectable concentration) of 0.016 milligrams per cubic meter (mg/m³), yet it is rarely present indoors in concentrations greater than 0.001 mg/m³ (though in one building whose occupants experienced symptoms, the concentration of octenol was almost eighty times the odor threshold). Experience suggests that the octenol concentration in air has to be more than ten times the odor threshold for it to be a human irritant, unless someone is chemically sensitive.

manufacturer has developed a mosquito trap that emits this odor; once the insects approach to investigate, they are sucked in and trapped. I read in one article that octenol is added to some perfumes as a fragrance. I wouldn't want to be wearing *that* scent in the woods!

There are researchers who believe that some symptoms experienced by people indoors are the result of exposures to MVOCs rather than to spores. While I believe that very sensitized individuals may suffer headaches as a result of exposure to strong mold odors, I think it's relatively rare for MVOCs to be present in high enough concentrations to cause many of the other respiratory symptoms people report experiencing.

Recent research done at the Monell Chemical Senses Center in Philadelphia, however, has shown that after repeated exposures, women of reproductive age can be more responsive to certain odors than men are. Because of some as yet unexplained genetic reason, women's odor thresholds (the lowest detectable concentration) for certain chemicals get lower with each in a series of exposures.* Thus, over time, women can become thousands of times more sensitive to

*Pamela Dalton, Nadine Doolittle, and Paul A. S. Breslin, "Gender-Specific Induction of Enhanced Sensitivity to Odors," *Nature Neuroscience* (Nature Publishing Group), vol. 5, no. 3 (March 2002).

certain odors than men can who are exposed to the same odors. And it has certainly been my experience that women notice musty odors more often than men do.

Investigators have tried to measure MVOC concentrations in the indoor air as an index for the presence of concealed mold growth. Unfortunately, many MVOCs are chemicals that may off-gas (change phase from solid or liquid to gas) from indoor sources other than mold, including perfumes and colognes. Detecting some of these chemicals in the air therefore doesn't necessarily mean that mold is present. In addition, many molds, even though they may be abundant in wall cavities, may not produce high levels of MVOCs. Since the presence (or absence) of MVOCs may not be a sure indication of mold (or the lack thereof) in an indoor environment, MVOC testing, which is costly, seems to me to be unnecessary.

Some people mistakenly identify the odor of bacterial growth (a sweat-sock odor) as a sign of the presence of mold. Bacteria usually grow in liquid environments, such as water or the fluids in plants or animal cells, and are associated with the odors of dirty sponges, rotting meat, dead animals, bad breath, and carpeting or towels that have remained damp too long. Though fungi and bacteria are very different organisms, they are both decomposers and thus compete for the same food. This competition is most likely why the fungus *Penicillium chrysogenum* inhibits the growth of bacteria, as the Scottish microbiologist Alexander Fleming observed in 1929. He was working with the bacterium *Staphylococcus aureus,* and he noticed that when *Penicillium chrysogenum* contaminated a petri dish, bacterial growth was absent in a ring around the mold colony. This discovery eventually led to one of the most important medical breakthroughs of the twentieth century: the development of the miracle antibiotic penicillin.

"Toxic" Molds

You have probably heard about the "toxic black mold" that has led people to evacuate and even demolish homes. The toxicity of a particular fungal colony is determined in part by the number and concentrations of mycotoxins (toxic chemicals that in most mold species are concentrated in the spores). Microfungi in the genera *Fusarium* and *Stachybotrys* produce the mycotoxins *trichothecenes*, which cause severe dermatitis (reactions such as blistering), nosebleeds, throat irritation, asthma symptoms, chest pains, and bronchial hemorrhaging. Various militaries have developed chemical warfare agents from trichothecenes, because less than 10 milligrams per kilogram of body mass is a lethal *ingested* dose for many animals, including human beings, and high inhaled doses can also result in shock and death.

The mold *Claviceps purpurea* (ergot) can produce several mycotoxins, some of which cause hallucinations (the hallucinogenic drug LSD is derived from ergot) while others cause constriction of peripheral blood vessels and lack of blood flow, thus leading to loss of fingers and toes due to gangrene. Even though ergot mycotoxins are dangerous, the vasoconstrictive ones have been used to treat migraine headaches, and the ones that cause muscle contractions have been used to assist birth.

The microfungus *Aspergillus flavus* produces mycotoxins called *aflatoxins*, which are lethal at high doses. They are also potent carcinogens and may cause liver cancer if eaten at low levels over a long period of time. The impact of exposure to aflatoxins by ingestion can vary drastically from organism to organism. If sensitive young animals regularly eat feed with 100 parts per billion (ppb) of aflatoxins, they will develop fatal liver cancer, though the effects may be minimal in older or less sensitive animals. Nonetheless, some regulatory authorities limit the permissible levels of aflatoxins in feed to a range of about 100 ppb up to approximately 300 ppb, depending on the animal.

Researchers have found that pregnant animals can pass ingested mycotoxins to their fetuses in utero. Scientists in France exposed pregnant rats to T2-toxin (trichothecene) and found that it passed easily through the placenta. In three other human studies, doctors from hospitals in the United Arab Emirates, Thailand, and Nigeria tested the umbilical cord blood after infants were born and found aflatoxins. Even at the low levels of mycotoxins found, the results of the studies suggested that the presence of aflatoxins was associated with lower birth weights.

Whether or not a mold produces a mycotoxin depends on many variables, including temperature, water availability, and the nature of the food source. Spores could be cultured from a sample of mold producing mycotoxins in a building and yet the new colonies might not produce high levels of mycotoxins in the petri dish. If the spores were grown under conditions identical to those in the building where the mold was first found, however, the mold might produce mycotoxins.

The type of mycotoxins a mold produces can also vary. Within a fungal species there appear to be different "subtypes" that produce entirely different families of mycotoxins. One subtype of *Stachybotrys chartarum* growing in a building may produce trichothecenes, while another subtype in a different building may produce a set of mycotoxins called *atranones*. The physiological effects of these mycotoxins are different, so toxic exposures to what is apparently the same black mold growing in two different buildings may yield different health consequences if mycotoxins are the cause. Finally, different genera of fungi may produce the same mycotoxins (for example, some species of *Fusarium* and *Stachybotrys* produce trichothecenes, and *Aspergillus ochraceous* and *Penicillium viridicatum* both produce ochratoxin).

It is correct to call a mold *toxic* only if it is producing mycotoxins, and this can only be determined by a laboratory through chemical testing of a sample of the dust containing the mold growth, or of the specific mold colony that produced the spores. (See table 3.)

Need We Worry?

Fungi are not always harmful. When you enjoy a piece of Camembert cheese or a buttery slice of sautéed mushroom, you are eating fungi. Sometimes you may not even know that you are eating mold unless you read labels. For example, I have recently seen the word *quorn* on the list of ingredients of some breads. Quorn is derived

Table 3. Fungi and Their Mycotoxins

Species	Mycotoxins
Aspergillus flavus	Aflatoxins
Aspergillus fumigatus	Fumitremorgens, fumitoxins
Aspergillus ochraceous	Ochratoxin
Chaetomium globosin	Chaetoglobosins
Fusarium graminearum	Zearalenone, vomitoxin
Penicillium expansum	Citrinin, patulin
Penicillium viridicatum	Ochratoxin
Stachybotrys chartarum	Trichothecenes, atranones, roridin

from the mycelium (hyphae) of a microfungus called *Fusarium graminearum*. Quorn seems to be growing in popularity with food processors because of its meatlike texture when converted into food products, though some people are worried about possible allergic reactions after ingestion. Fungi can also make contributions to medicine. The antibiotic penicillin, as noted earlier, and the immune system suppressant cyclosporin, required for organ transplants, are products of fungi.

You often hear the argument that mold spores are everywhere. People have been raking leaves and tossing moldy hay for centuries; they didn't worry about mold hundreds of years ago, so why should we worry now? Many of the spores found in outdoor air are associated with macrofungi and with crop or plant leaf diseases. In my experience, exposures to such spores tend to be less of a health problem than exposures to the spores from the microfungi associated with decay in soil. In addition, the production of worrisome mycotoxins is more common among soil-inhabiting microfungi. Unfortunately, these fungi (such as *Stachybotrys chartarum, Aspergillus fumigatus,* and *A. versicolor*) grow more commonly indoors (in "sick" or chronically damp buildings) than outdoor-type fungi do. For example, a recent journal article reported that ochratoxin was found in dust collected from moldy carpets (one carpet contained this mycotoxin at a level of 20 ppb). In one home that I investigated, the moldy dust from a duct contained 1,500 ppb of this mycotoxin.* When consumed by humans, ochratoxin causes kidney disease, suppresses the immune system, and may be carcinogenic. It is also possible that the combined presence indoors of several species of mycotoxin-producing molds could have a greater effect on someone's health than the effect of any one of these fungi alone.

Spores

The size of the spores matters when it comes to potential exposure. Bigger spores settle out of the air more quickly, while smaller ones remain airborne (suspended in air) for minutes to hours. For example, *Stachybotrys* spores are relatively large and are produced in clumps, stuck together by mucilage. They thus are rarely found in the air, whereas *Aspergillus* and *Penicillium* spores are very readily aerosolized (dispersed into the air), because they are small and grow in fragile chains. Smaller spores can penetrate more deeply into the respiratory system, and therefore they produce higher exposures, whereas larger spores, or clumps of spores, will be trapped in the nose and on other surfaces of the upper respiratory system. Still, while spores trapped in the nose remain there, spores that enter the lungs may be swallowed, because they are eventually carried out into the mouth by the fine hairs (called *cilia*) that line the respiratory system. And even if the spores are no longer viable (alive), they can still contain allergens and mycotoxins that may affect health. (Some of the mycotoxins and toxins associated with other microbes that grow alongside the mold paralyze the cilia and therefore make it more difficult for the human respiratory system to clear inhaled particles.)

---

*John L. Richard, Ronald D. Plattner, Jeff May, and Sandra L. Liska, "The Occurrence of Ochratoxin A in Dust Collected from a Problem Household," *Mycopathologia* (Kluwer Academic Publishers), vol. 146 (1999):99–103.

Finally, we must consider an unknown factor: the amount of mycotoxin on or in the aerosolized fragments of spores (or the hyphae, as they too may contain mycotoxins), or on airborne particles that have been in intimate contact with living mold colonies or toxic spores in dust. Surfaces don't have to be in direct contact with mold growth to contain mycotoxins, however. Though most mycotoxins are insoluble (do not dissolve) in water, some of the insoluble mycotoxin in *Stachybotrys chartarum* is bound to water-soluble components and can therefore migrate through porous solids by capillary action.* This process might explain why mycotoxins are sometimes found in some seemingly mold-free inorganic building materials just beneath surfaces, where microfungi are or have been growing.

Recently, researchers have found that when some mold colonies are disturbed, large numbers of very small particles—approximately

*Information received in a personal communication with Bruce Jarvis, an organic chemist at the University of Maryland.

Aspergillus species are very common in indoor and outdoor air. When species of this microfungus (often *A. fumigatus*) grow in lung tissue, however, an occupational disease called *aspergillosis* can develop. This disease is often diagnosed in workers exposed to high levels of the fungal spores. Warm environments such as compost piles and even birds' nests can be sources of hazardous spore levels if the moldy sources are disturbed. Nests containing *A. fumigatus* are sometimes found at the rooftop intakes for hospital ventilation systems, leading to the possibility of patients being exposed to spores. The splashing of water in showers with *Aspergillus* mold growth (possibly in drains) has even resulted in cases of aspergillosis in patients who have compromised immune systems and who are being treated in a hospital's bone marrow transplant unit.

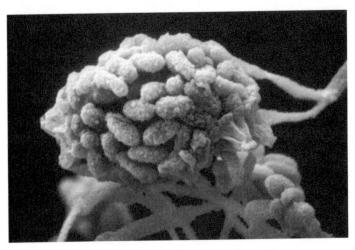

A clump of *Stachybotrys* spores. This desiccated clump is from an evaporative (cool-mist) humidifier (see chapter 6). The spores are grouped together, still stuck to each other and to the conidiophore upon which they formed. The evaporative pad that the mold grew on is made of paper (cellulose) and was not replaced soon enough. This mold grows best on constantly wet cellulose. Whose idea was it to make this pad (called a wicking filter) from paper? (1,400x SEM)

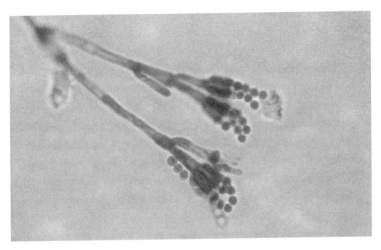

Penicillium conidiophore in the air. This branched structure came from a sample of air taken in a wet basement. The mold was growing on the unpainted paper of a drywall ceiling and became airborne after the surface was disturbed by a breeze. The spheres are the spores (each about 3 millionths of a meter in diameter) that grow in chains. (1,000x light)

one-tenth the size of a spore—are released. These particles, called *micro-particles,* can come from the surfaces of spores, or from hyphal fragments or crystals from mold metabolites or building materials. It has been postulated that these micro-particles may carry allergens as well as mycotoxins, and because they are so small, they are not trapped by the nose and they travel deep into the lung.

All that said, I believe that cause-and-effect relationships can sometimes be exaggerated. I have seen it stated on the Internet more than once that a single fly could carry over seven billion *Stachybotrys* spores, estimated by the author to be a lethal inhaled dose for over one hundred thousand individuals. Yet it takes 100 billionths of a gram of instilled mycotoxins, or more than a million toxic *Stachybotrys* spores, to kill a mouse. And a person might be a thousand times heavier than a mouse! In addition, seven billion spores would weigh about two-thirds as much as a penny and have about the area of a postage stamp—an impossible load for a fly to carry.

It is also my opinion that there is not enough mycotoxin in an inhaled spore or even several inhaled spores to cause the kind of immediate toxic reactions that some people fear. I do believe, though, that in highly sensitized individuals, immediate allergic responses can occur after inhalation of only a few spores. (For comparison, consider that 20 billionths of a gram of allergenic enzyme, the amount found in a single dust mite fecal pellet—about the size of a mold spore—is sufficient to cause an allergic reaction during inhalation or in a skin prick test.) And after chronic inhaled spore exposures, some individuals say they experience memory loss, headaches, joint pain, and fatigue. These symptoms pose serious concerns, and a great deal more research is needed.

Mold in the News

Given the level of concern about mold, it's no surprise that mold is a frequent topic in the news. In December 1999 an article called "Mold: A Health Alert," by Arnold Mann, appeared in *USA Weekend.* The story and image of Melinda Ballard and her husband, Ron Allison, wearing protective suits and masks in their dream house awak-

The amount of mycotoxin in a toxic *Stachybotrys chartarum* spore might only be about 0.1 picogram or 0.0001 trillionths of a gram (Bruce Jarvis, personal communication), and the number of spores in an outrageously contaminated space might be 1,000 spores per cubic meter. Since there are 1,000 liters in a cubic meter of air, this corresponds to a concentration of one spore per liter of air. We breathe about eighteen times a minute, with each breath taking in about half a liter of air, so breathing air for one minute with this improbably high concentration of spores would result in a maximum of nine spores being deposited in the respiratory tract (or about 0.9 picograms of mycotoxin).

It takes over a million toxic *S. chartarum* spores instilled into the nasal passages of a mouse to kill it, and a mouse's weight is only a small fraction of a human's weight. Breathing in the mycotoxin from nine spores of this microfungus should therefore not have much of an effect on human health (excluding any allergic reactions sensitized individuals who breathe in these nine spores may have). Some researchers hypothesize, however, that when spores land on lung cells, mycotoxins are released in high concentrations at that microscopic site, and the toxic effect on those particular cells is heightened.

ened the nation to the hazard of living in the midst of the mold *Stachybotrys chartarum*. "I call it the 'house of pain,' said Ballard."

After plumbing leaks soaked and warped their hardwood floors, the family suffered headaches, fatigue, and respiratory and sinus problems, and they eventually fled the house. Claiming that mold grew because Farmers Insurance had refused to authorize the timely repair of the leak, Melinda and Ron brought suit and were awarded $32 million in damages. (On appeal, the award was reduced to $4 million.)

In another mold case, a family developed asthma, allergies, rashes, and other medical problems after moving into an apartment in Anaheim, California. Long-term water intrusion had led to concealed mold growth, and a press release from a law firm in California announced a $900,000 "toxic mold" settlement.

Some people believe that the media hype is fueling exaggerated concern. One environmental investigator, commenting on the calls he received after TV coverage, said, "We call them 'nut calls.'"* Another consultant who has been a defense witness in mold cases believes that people's worries are excessive, and that "a great deal of money is being wasted on this issue."†

*Ellen Barry, "Air Concerns Grow along with Mold," *Boston Globe,* September 2, 2001.

†Julie Deardorff, "Mold Fear Growing, but Data Lacking," *Chicago Tribune,* May 26, 2002.

After investigating several hundred buildings in which inhabitants suffered nonspecific symptoms such as headaches, fatigue, and eye and throat irritation, the National Institute for Occupational Safety and Health (NIOSH) initially concluded that symptoms associated with "sick building syndrome" (SBS) were largely the result of an inadequate supply of fresh air in buildings, and that at most only about 5 percent of SBS was due to microbial (including mold) exposure. This conclusion was based on investigations in which there was not much emphasis on testing for microbes. After additional and more comprehensive investigations, NIOSH now suggests that from 35 to 50 percent of SBS is due to microbial exposures. I suspect that the percentages will increase again as new cases of SBS are investigated more thoroughly, using instruments and techniques that can detect the presence of microbes and the products of their growth indoors.

Since Arnold Mann's article appeared, though, the number of insurance claims has increased in some states by a factor of ten. It's thus no surprise that in California, Connecticut, Florida, New Hampshire, and South Carolina, among other states, insurance companies are taking steps like refusing to sell policies to owners whose homes have suffered recent water damage, or denying claims for mold damage caused by poor home maintenance. And pointing to heavy losses, insurance companies are asking for hefty increases in premiums.

I have a lot of sympathy for people who have suffered because of mold in their homes. In fact, I believe that the enormous impact mold can have on health is still not adequately recognized. On the other hand, I have investigated several mold claims for insurance companies and found that the homeowners ascribed damage to mold growth that had clearly been caused by inadequate maintenance over the course of many years. I think such misrepresentation unnecessarily increases the cost of insurance and fuels the skepticism that doubters express about the dangers of mold.

The growing number of claims is threatening more than just insurance companies and homeowners. Landlords are also increasingly vulnerable. I have heard that some are even adding language to lease agreements asking tenants to recognize that the premises are mold free and to promise that they will continue to maintain the premises in that condition. Thus, the responsibility for mold problems is transferred from the landlord to the tenant.

Builders are also finding themselves in a potentially precarious spot, because they may be held liable for mold problems that develop in new homes. Yet mold growth is a common occurrence in construction. "In a survey by the NAHB (National Association of Home Builders), 28% of builders nationwide reported having mold in at least one house under construction last year, and 18% said the problem had appeared in occupied houses."*

*Steve Kerch, "Breaking the Mold," CBS.MarketWatch.com, March 8, 2002.

Mold isn't bad news for everyone. "Mold is gold" is an expression I have heard a lot lately, and the number of mold testing labs and remediation companies is increasing. Many of these companies are committed to delivering honest services, but I have also heard of some disreputable remediation practices. In Texas it's rumored that a mold mitigator misted the inside of a house with water and sealed it up for a few weeks, thus incubating a small mold patch into a mold catastrophe. (I think this practice is called *house cooking.*)

I have also read about a Texas developer who purchases mold-damaged properties for less than 50 percent of their fair market

Concealed mold. There was a musty odor in this room, so the homeowner pulled the paneling away from the drywall in his just-purchased house to investigate. At the left, you can see that the back of the paneling was covered with white mold, fuzziest at the bottom. At the right, black *Chaetomium* mold was growing on the back corner of the drywall paper where it touched the floor. This dark mold most often grows in soaked cellulose. This room was only one of many areas where there was concealed mold growth caused by roof leaks.

value and then renovates them for resale. I only hope that the mold problems were properly remediated. I inspected one home in Massachusetts in which mold-damaged walls, floors, and ceilings may have been covered over to conceal the growth. The new owner, who has mold allergies, moved in and became ill almost immediately. The last I heard, the family was planning to sue the real estate agent and demolish the house.

Are Guidelines an Answer?

The mold landscape is in chaos, so it makes sense that there are movements in Arizona, California, Connecticut, Texas, and Indiana, among other states, to protect homeowners and stake holders. In California, for example, the Toxic Mold Protection Act, effective January 1, 2001, requires disclosure of current or past mold growth. And in Arizona the building industry supported a bill that made it harder for a homeowner to sue a builder.

On the federal level, Michigan Congressman John Conyers, Jr., introduced the Melina Bill, named after an eight-year-old girl who developed asthma after being exposed to mold in her home. The bill includes a requirement for research, standards and licensing for mold remediators, changes in construction practices, inspections for mold, and insurance pools for those who have suffered from exposure.

One way to protect the public health would be to establish guidelines for indoor mold spore concentrations. Unfortunately, it will not be easy to set such limits the way they have been set for chemicals, lead, or asbestos, because mold is naturally present outside, and spores can easily find their way into a building. Nonetheless, I applaud efforts that encourage a scientific understanding of mold and that foster standards of maintenance to reduce indoor mold growth. I do think, though, that for the immediate future, sights should be set on attainable goals: no visible mold growth on walls and ceilings, no mold growth in carpeting, and adequate filtration in hot-air heating systems and air-conditioning equipment to minimize the accumulation of the dust that serves as food for microbial growth.

In the end, whether you are a homeowner or condominium owner, a tenant, a property manager, or a landlord, staying abreast of the news in medicine, legislation, and the insurance industry will help you defend your physical and economic health against mold.

PART II

THE SEARCH FOR MOLD

To eliminate mold, you have to know where it's located. Look carefully at the obvious surfaces: ceilings, walls, and floors, particularly in areas where there are stains or where there has been leakage. Also check in places where you would expect the surface temperatures to be low: on "outside" ceilings and walls (particularly in cold closets) and near windows. To make it easier to see the mold colonies, hold a very bright flashlight parallel and close to the surface in question. And don't forget to look behind bookcases (at the backs and at the walls), at the sides and undersides of furniture (a mirror is helpful), and along baseboards.

Cupboard with mildew growth. In some cases mold growth on furniture is obvious, and other times it's not. This piece of furniture was stored in a damp basement.

If you are allergic to mold, wear a mask with at least a NIOSH (National Institute for Occupational Safety and Health) rating of N95. (A disposable mask with this rating always has two straps and is thicker than the inexpensive single-strap mask used for nuisance dust.) Take care not to disturb mold on surfaces or in dust, for even if you aren't allergic or sensitive to mold, others in your home may be. If you enter a very moldy basement, don't wear the same clothes or shoes back into the rooms of your home. (Clean the clothing by normal washing.) You may want to have someone else do the looking (possibly hiring someone), particularly if you have mold sensitivities or if you want a moldy crawl space investigated. But again, that person should be protected and should take care not to unnecessarily disturb mold growth or moldy dust.

Unfortunately, sometimes mold growth isn't visible, and the sources of contaminants can be right under our noses. Mold growing unseen in a stairway rug or a puffy cushion is likely to be a problem, whereas visible attic mold, unless disturbed, is much less likely to be an issue. If you have mold sensitivities, try to keep track of where and when you experience symptoms in your home. If wearing a NIOSH N95 mask, which filters out nearly all of the particulate allergens, helps reduce your symptoms, then airborne particulates are most likely an issue for you. The next step is to find the sources of the particulates and eliminate them.

It may be helpful to narrow the list of possible sources of mold contaminants. If there are rugs or window drapes in the room, have someone roll them up carefully (to minimize the spread of potentially moldy dust) and wrap them in plastic, which prevents the release of particles. (Refer to the description of containment-like conditions in chapter 10 before you begin. If anyone in your household has mold sensitivities, *use extreme caution*, even when just removing rugs from your home.) If there are couches and chairs with cushions, these too can be temporarily wrapped in plastic and sealed with duct tape.

Wall-to-wall carpets can be temporarily covered with a heavy drop cloth. Don't use thin or slippery plastic. Use a drop cloth that has plastic on the underside and cloth or a clothlike material on the top surface. Don't leave slippery surfaces or uneven edges that people can trip over. These steps may seem drastic, but in a way, you are a detective involved in solving a mystery, and part of your task is to eliminate suspects. If covering everything makes no difference when the heat or air conditioning is on, then there may be mold growth in the heating or cooling system, which may have to be professionally cleaned. (Turn the system off, air the house out well for ten minutes, and see if that makes a difference.)

If your search does not help you identify the location of mold growth and yet your home still smells musty or you continue to experience allergic symptoms that may be mold related, you should probably consider hiring a professional mold investigator.

Chapter 5

WHAT LURKS BELOW

Does your finished basement contain a play or exercise room? Is there a laundry area in your unfinished basement? Perhaps you have a crawl space under a family-room extension. Like it or not, a below-grade space, whether finished or unfinished, is still a part of the house, because 30 to 50 percent or more of the air in your bedroom or living room can come from the basement or crawl space beneath. If your home has hot-air heat or central air conditioning, most of the air in the habitable rooms can come from the lower level if there is a basement return or a disconnected return duct in a crawl space.

Particularly in homes with allergic family members, *a below-grade space should not be maintained any differently than the rest of your house.*

Crawl Spaces and Unfinished Basements

Crawl spaces and unfinished basements should be kept clean and well lit so you can investigate, if necessary, to check for moisture problems or mold growth.

Crawl Spaces

In one e-mail I received, a homeowner said, "My crawl space stinks, but I'm afraid to go down there because it's dark." The man's fear of his dark and smelly crawl space explains in part why crawl spaces can be such a problem. Even if intermittent, a musty smell in a crawl space is a sign of mold growth. (Some molds produce microbial volatile organic compounds, or MVOCs, when they are actively growing in humid weather.) MVOCs can diffuse through the flooring or infiltrate with spores through any openings around pipes or ducts passing from or through a crawl space or basement into the rooms above.

One couple contacted me because they had bought a house built over a damp crawl space. When they moved in, there was visible mold on the floor sheathing and joists, and much of the wood was decayed. They replaced the rotted wood, increased the ventilation, sprayed the mold with a bleach solution, and added gutters to the house. Still, in humid weather the rooms above the crawl space had a musty smell.

Grade means the level of the ground. A house built on a concrete slab poured on the soil is called *slab on grade.* If a home is built nestled into a hill, the front of the lower level with an exit door is *walk out to grade* and the back is *below grade.* Some homes are built partially or entirely on crawl spaces, which can be either below or above grade. Some houses have full foundations set deep into the soil, and the basement space is entirely below grade. Some foundations are exposed at the top of the walls, usually with windows installed, and then the basement is partially below grade.

Whether below or partially below grade, a basement can be unfinished (exposed stone, block, or concrete walls and concrete floor) or finished (drywall or paneled walls with ceiling and carpeting). In larger residential buildings, some basements are finished into habitable units, either below or partially below grade.

Their situation highlights several important issues about dealing with mold in crawl spaces. First, if macrofungi are present, it's extremely important to replace as much of the severely decayed wood as is feasible; otherwise, should the relative humidity increase sufficiently or the wood get wet, any fungus that is living will spread to any newly installed, unprotected wood that is in contact with the decay. And second, to minimize future growth of microfungi, rather than spraying a crawl space with bleach solution, I recommend that the framing, after being professionally cleaned (if needed), be treated by a PCO (pest control operator) with a *borate* mildewcide. This leaves a residue to protect the wood. After the mildewcide dries, a light coat of spray paint can be applied to the entire wood structure (joists, subflooring) to seal in any loose mold spores. When water-based materials are used, the crawl space must be allowed to dry out before being closed up. If the weather is dry, use an exhaust fan. If the weather is humid, use a dehumidifier. If solvent-based compounds are used, the crawl space should be ventilated at least until the odor is gone before being closed up.

Ironically, permanently increasing the ventilation in a crawl space may actually worsen a mold problem by introducing more humid air. If you live in an area where the relative humidity is regularly or seasonally high, I recommend sealing a crawl space from the exterior and using dehumidification to keep the area dry (relative humidity at no more than 50 percent). Building codes in many regions mistakenly (in my opinion) require crawl space ventilation. Many homes with crawl spaces that I have investigated had mold problems, caused in part, I believe, by the high relative humidity of the ventilated crawl space itself. Unfortunately, even if you do understand how to take care of a crawl space, the previous tenant or owner may not have, and this is part of the difficulty. If a member of your family has allergies or asthma, or if you are nervous about entering and maintaining a crawl space, I advise you to avoid living in a house with a moldy crawl space.

Unfinished Basements

In an unfinished basement, items leaning up against the foundation walls or resting directly on the floor are cooled in warm weather by the concrete. Moisture can then condense on these surfaces if they are below the dew point of the air, and microorganisms will grow. I often find mold growth on the bottoms of cardboard boxes resting on a basement floor. Sometimes, if the floor has been very damp, when you lift the box its imprint will be left in mold, or the bottom of the box will be so deteriorated that it separates and sticks to the concrete.

Microfungi also often grow on the legs and undersides of wooden furniture, and in the cushions of upholstered furniture, if the pieces have been stored in a damp, unfinished basement space.

Floors

Dirt floors in crawl spaces and unfinished basements should always be covered. If you have a dirt floor, the minimum step is to completely cover the soil with heavy-duty vapor-barrier (retarder) plastic, intended for such use. It's a good idea to eliminate all debris under containment conditions (isolated from the rest of the house; see part 3) before installing the plastic, and to place mildew-resistant or pest-resistant planks on top of the plastic to allow access to mechanical equipment or pipes. (And in my opinion, no one should enter or work in a dirt-floored crawl space without wearing respiratory protection.) You may see water droplets under the plastic, which is telling you that the vapor retarder is doing its job. You should *never* see water on top of the plastic, however, because that indicates a problem: either a plumbing or foundation leak, or condensation due to excess relative humidity that must still be controlled.

A concrete floor is preferable to a vapor retarder alone. When installing a concrete floor in a basement or crawl space, you will most likely have to remove some soil first in order to accommodate the crushed stone that goes under the concrete. Typically, builders place a plastic vapor retarder on top of the crushed stone before pouring

Mildew on the bottom of a table. The white spots are mildew (mold) growing on the bottom of an antique table that no doubt was once stored in a damp basement. Furniture like this can be a source of musty odor in a seemingly pristine home. If the mold is present on the bottom drawer of a dresser, spores are aerosolized every time the drawer is opened or closed.

concrete. Soil particles are very small and promote capillary action, so placing crushed stone with a plastic vapor retarder between the soil and the concrete minimizes the movement of moisture upward. Painting the concrete after curing (drying) acts as a further guard against moisture diffusion. (Remember to ventilate or dehumidify a space where fresh concrete has been poured.)

In homes built before the 1960s, it's unlikely that there is plastic under the basement floor, so painting the concrete can reduce the diffusion and evaporation of water. In fact, I believe that painting foundation walls and floors is an important component of basement hygiene. The surface of concrete is full of pores that can never be completely cleaned; sealing the concrete with paint partially fills the pores and makes cleaning easier. Many manufacturers make paints that are specifically for use on concrete. But please note: in a basement or crawl space, I would definitely avoid using any paint that contains butyltin or other fungicides intended for outdoor use, as some of these products can off-gas and create odor problems indoors.

Finished Basements

I've seen beautifully decorated basement playrooms, studies, exercise rooms, bedrooms, and even theaters that were infested with concealed microfungi. Many of these rooms were designed for leisure activities, so people spent hours there, breathing in spores. If you exercise in a moldy basement room, your exposure is even greater because of your increased respiration rate. (It's ironic that the areas where we go for recreation are often the most contaminated spaces in the house.)

In finished below-grade rooms, carpeting and the pad beneath that rest on the concrete floor can be soaked by floor flooding or dampened by water diffusing through the concrete (if not sealed by resilient tile or vinyl), or by water vapor from the air (if the carpet or pad resting on the cooler concrete is intermittently below the dew point). This leads to mold and mite infestations. Finished walls, particularly near the concrete floor, are also cooled by heat loss to the foundation, and they too can become colonized. Microfungi can grow on dust, often in an unseen film on basement wall paneling and the unfinished surfaces of wooden furniture, and hidden in cushions (especially where food was spilled), upholstery, and carpeting. And if the relative humidity is high, white acoustical ceiling tiles can occasionally be covered with barely visible mold colonies. When this happens, as with any unseen fungal growth, the room can have a ubiquitous musty smell, even when the area looks clean.

Should You Avoid Finished Below-Grade Spaces?

As far as I am concerned, people who have allergies and who cannot rigorously control indoor relative humidity should avoid living in homes with finished, carpeted below-grade spaces. When people with mold allergies call me to ask if they should finish their basements into playrooms or offices to gain more space, I discourage them. Better to move into a bigger house than to set up conditions in your home that increase the risk of mold growth. (Georgia-Pacific manufactures a relatively new wall covering called DensArmor Plus™, which may be more appropriate for below-grade walls. In this composite, the gypsum is sandwiched between two layers of fiberglass rather than paper. As long as the fiberglass is clean—no biodegradable dust—and painted, it cannot serve as food for mold the way paper on drywall can.)

I also believe that people who have sensitivities to mold should be very cautious about living in slab-on-grade houses with carpeting on the lower level, or in carpeted basement apartments. A young man called me because he was concerned about the sinus infections and headaches that his bride was experiencing. After they got married, she moved into the basement apartment where he had been living for several years. She was a consultant and spent most days traveling. She felt fine when she was away but began to feel ill within hours after returning home.

The man had maintained the apartment very well when he was single, though he had complained to his landlord twice about flooding that had wet the carpeting in the bedrooms and the closets. I took samples of the air and dust and found that they contained the same kind of mold spores. The carpet was clearly contaminated, and whenever anyone walked on its surface, spores became airborne. Dust on the couple's clothing in the closets was also full of the spores.

The couple decided to move, but to reduce their exposure to mold while they were looking for a new home, I recommended that they immediately cover the carpet with an impervious material (such as a drop cloth or nonslip plastic sheets), eliminate one couch that contained mold growth, seal another couch in plastic, and have their clothing washed or dry-cleaned, with the proviso that any clothing with substantial visible mold growth be discarded. After the couple took these steps, the woman was able to spend time in the apartment, free of symptoms. Ultimately the source of the leaks had to be repaired, the carpeting removed (and hopefully replaced with hardwood floors or ceramic or resilient tile), and the apartment air-conditioned or dehumidified.

Insulation in Finished and Unfinished Below-Grade Spaces

Although I don't usually see mold problems with fiberglass in attics and in above-grade walls (unless there has been a pest infestation or leakage in a wall cavity), it's my opinion that in finished basements, it's always preferable to insulate the inside of the foundation walls with one- or two-inch foil-coated sheet-foam insulation, rather than to use fiberglass insulation between the wall studs. Sheet-foam insulation reduces the heat loss from the wall framing and minimizes the likelihood of dew point conditions. In addition, foam is not porous the way fiberglass is and thus will not be saturated by a floor flood. (If you can afford to lose the room space, I also strongly recommend leaving enough distance between the stud wall and the insulated foundation so that you can walk through to clean and inspect the space.)

In a finished or unfinished basement, if you don't want hard, cold floors you can install plywood over two-inch foil-coated sheet-foam insulation (but check with the manufacturer and a contractor first on the load capacity of the foam). This will give you a warmer floor while minimizing the likelihood of mold growth in the wood. If you insist on having solid wooden shelves in an unfinished basement, rather than the open metal or plastic ones that I recommend, and if the shelves are placed up near the foundation wall, it makes sense to place sheet-foam insulation between the backs of the shelves and the foundation. This will also help prevent condensation and mold growth on the wood surfaces and on boxes stored close to the foundation wall. Keep in mind that some building codes have restrictions on the use of foam insulation because this material can produce toxic fumes in the event of a fire. In some cases the foam must be protected with a layer of more heat-stable wall material such as drywall or panel.

Exposed fiberglass insulation in a basement ceiling can incubate microfungi (typically species of *Penicillium, Aspergillus,* and *Cladosporium*), particularly if someone has been cutting wood in a basement and sawdust has accumulated on the fibers (remember, some microfungi can grow without the presence of liquid water, if the relative humidity is over 75 percent). Because I've seen so many mold problems with open ceiling insulation, I do not believe that fibrous insulation should be placed between basement or crawl space ceiling joists and left exposed, unless you dehumidify. I have one other concern about fiberglass insulation in crawl spaces: rodent infestations. Mice and other field rodents love to nest in fiberglass, which leads to odor and mold problems.

Some manufacturers make fiberglass batts encased in plastic, and this should prevent dust accumulation in the fibers. I haven't seen any installations of this material yet, so I don't know if there are any problems associated with the use of these batts. If you decide to use this type of product, check to make sure that the plastic contains fire

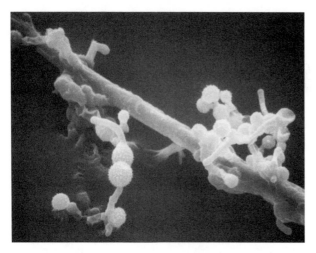

Microfungi wrapped around fiberglass. The diagonal tube is a fiberglass fiber from insulation in a crawl space. It is entangled by hyphae with spores. There are no nutrients in glass on which the mold can live (glass is a mineral). The mold is thriving on the house dust trapped within the insulation. (1,700x SEM)

retardant. If you already have exposed fiberglass insulation, do not cover it with any combustible plastic sheet material, such as polyethylene.

Mold-contaminated or rodent-contaminated exposed fiberglass insulation should be removed professionally under containment conditions (see part 3) and the joists and subfloor (or framing) lightly spray-painted to contain potentially irritating or moldy dust.

Dehumidifiers

Since mold can grow wherever there is dust and moisture, it's important to keep the relative humidity (RH) in below-grade spaces at no more than 50 percent. One family ran two dehumidifiers in a finished basement, and yet the house still smelled musty. It's always a good idea to dehumidify a basement, whether finished or unfinished, but be sure the machine you use is up to the task. You might consider using a larger-capacity dehumidifier, such as those supplied by Therma-Stor® (see Resource Guide at the end of the book). Even if you have adequate dehumidification, conditions of high RH, conducive to mold growth, can still develop in rooms that are shut off from the rest of the space. So keep doors open or use more than one dehumidifier, and measure the RH with a thermo-hygrometer.

You can place a dehumidifier over a sink or floor drain so you don't have to keep emptying the bucket. Your other option is to use a condensate pump (available from building supply or hardware stores) to collect the water and pump it outside or into a sink or washer drainpipe. Finally, don't forget to keep the dehumidifier coils clean (and use a dehumidifier that has a filter for the cooling coil).

Sources of Moisture

What are some of the sources of moisture that can lead to high relative humidity, condensation, and mold growth in below-grade spaces? Gallons of water can evaporate daily from a dirt floor, and moisture can diffuse (though much more slowly) through an uncovered concrete floor or foundation. Exterior summer air that contains moisture can infiltrate through leaky basement windows and doors (including the bulkhead) and through crawl space vents. Condensation will occur when the concrete foundation walls and floor, as well as the surfaces of cold-water pipes and uninsulated air-conditioning ducts, are below the dew point of the air. I have even seen water dripping from soaked fiberglass ceiling insulation, due strictly to condensation of moisture from air on the colder batts after the weather suddenly changed from cool and dry to hot and humid.

If the yard is not properly graded, rainwater can also enter the lower level from the exterior soil and lead to conditions of high relativity humidity and to mold growth. It's thus important to have good grading and drainage around the foundation. If the ground outside is graded toward the foundation, or if the gutter system is clogged, water can pool along the outside of the house and enter the basement through foundation or floor cracks or the floor/wall joints. Roof water should be directed away from the foundation, with either downspouts or an ample roof overhang. In most cases it makes more sense to stop the liquid water from flowing into a below-grade space than to try to remove the moisture with a dehumidifier after it evaporates within.

The level of the water table (the level below which the soil is saturated with ground water) can also cause trouble. If the water table in an area is naturally high or becomes elevated during a heavy rain, water can enter the foundation through wall or floor cracks. A house built in an area with a high water table may have a sump pump installed in the basement, to pump water out from beneath the slab and thus to prevent floor flooding. Sometimes people who are house-hunting are worried if they find a sump pump in a basement. I wouldn't automatically rule out such a home, because many homeowners install sump pumps out of caution. If a sump is full of water, however, there is reason for concern, because this may indicate a high water table. In addition, any stains on stored goods and the walls in the basement indicate flooding—even more of a reason for worry.

Laundry areas can also cause water and mold problems in crawl spaces beneath. In one newly purchased home the washing machine was draining into the crawl space under the master bedroom. Even though the owners repaired the washing machine, mold had already started to grow. In another home, a side-by-side two-family residence, the tenant's apartment was located over a crawl space. She was irate because of a powerful musty smell in her unit and was threatening to sue her landlord, who lived next door. I discovered that the owner's clothes dryer vented hot, moist air directly into the insulated ceiling joist bay in the crawl space. The moist insulation was filled with mold flourishing on the accumulated lint.

Of course, it's also important in a basement or crawl space, or in the lower level of a home built slab-on-grade, to eliminate water from a plumbing leak, or from a burst hot-water tank or washing machine hose as soon as possible. Professionals are usually required for clean-up and repair after below-grade flooding, whether from a broken pipe, high water table, water entry from the outside, or sewer backup (always call professionals when sewage is involved; see part 3).

The Microscopic Food Chain

Mold attracts foragers in the microscopic food chain of life. In basements I often find that most of the mold present has been partially digested and excreted by insects, such as mold-eating mites. When samples are viewed with a microscope, what appears at first glance to be a wall covered with mold and spores can prove to be a wall covered with insect fecal pellets (waste products) containing spores. Thus, the sensitized are potentially exposed to both mold and mite allergens. Spiders can also be an indication of excess moisture and mold growth, because spiders only feed on live insects. If your basement is full of spider webs, there's a good chance that an array of insects are feeding on each other and, ultimately, at the bottom of the food chain, on microfungi.

Efflorescence Is Not Mold

Some concerned homeowners have described large tufts of fine white material resembling a fungal mycelium on foundation walls, on brick support columns for beams, and at the joints between vinyl

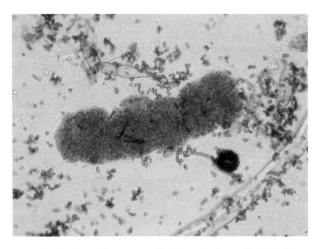

Insect fecal pellet with spores and hyphae. The dark cylindrical object at the center is an insect fecal pellet from moldy paper on a basement floor. Within the pellet you can see the spores and some of the hyphae that the insect ate but barely digested. All around the pellet are individual spores from the mold colony. (200x light)

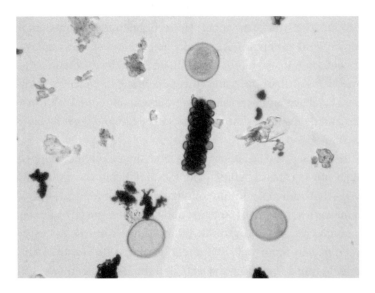

Insect fecal pellet with *Chaetomium* spores. The three light-colored round objects forming a triangle are plant pollen. Within the triangle is a dark 63-micron-long cylindrically shaped insect fecal pellet containing partially digested *Chaetomium* spores. Pollen, insect fecal pellets, and spores are just some of the materials that make up house dust. (200x light)

or asphalt (resilient) floor tiles in basements. Some of these tufts appear to be snowy-white growths arising from crevices. These tufts will actually push large sections of paint off the foundation wall and then accumulate in piles on the floor. When the tufts appear between floor tile joints, the edges and corners of the tiles may be forced upward and broken.

Liquid water that is pure (such as distilled water, which contains no dissolved minerals) leaves no residue when it evaporates. When water containing dissolved minerals evaporates, however, the minerals are left behind as crystals. As moisture moves through masonry or plaster walls, the water dissolves some of the minerals in the concrete or plaster. The mineral crystals that are left behind when this water evaporates are called *efflorescence.*

One morning a teary homeowner contacted us in an absolute panic. Two days before, a heating contractor had gone into her basement to check her furnace and had told her that the white fuzz on the foundation walls was mold. He recommended a remediation company, and a representative took one look and gave her an estimate of $3,000 to clean the "mold." In addition, he encouraged her to move out of the house until the work was completed. She had taken his advice and was calling us from a hotel room. We investigated and determined that all of the so-called mold growth was in fact efflorescence.

Efflorescence on basement walls and floors can be reduced, though not necessarily eliminated, by minimizing the amount of outside moisture close to the foundation. A drier foundation wall also means less basement moisture and humidity—an environment less conducive to mold growth. Efflorescence can also occur on kitchen or bathroom walls that are in contact with slowly leaking plumbing pipes. Repairing the source of the leakage can solve the problem.

If you have tufts of what looks like efflorescence on the concrete of your basement floor or walls, scoop up some of the material and see if it dries out. (I have taken a cup full of these "hydrated" crystals out of a basement and seen them shrink into a gray, dusty pile as the

White "growth" due to mineral efflorescence. This picture gives us several pieces of information. First, the narrow vertical patch extending downward from the corner of the window is a foundation crack that has been patched on the inside only. Second, the horizontal white staining pattern beneath the window indicates the level of the grade (ground), where rainwater is accumulating and moving through the foundation by capillary action. The white vertical patterns at either side of the vertical foundation crack illustrate that water is entering the crack below grade and moving both horizontally and vertically by capillary action (as well as downward by gravity). These staining patterns are caused by efflorescence.

moisture trapped in the crystals evaporated.) If you are not allergic to mold, you can also collect a sample of the material (carefully, wearing a NIOSH N95 mask) and put a few drops of vinegar on it. If some of the crystals bubble and some disappear, the material is definitely mineral and not mold growth. If the white material is hairy and none of it dissolves, it's probably fungal hyphae.

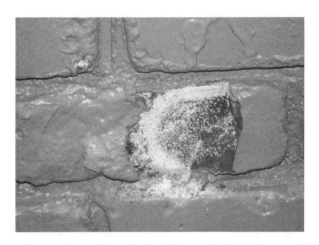

Efflorescence crystals on a brick. A thick coat of paint had been applied to this below-grade brick wall in an attempt to seal it. Nonetheless, water moved through the masonry by capillary action. The resulting growth of crystals (efflorescence) under the paint surface pushed off the paint. The owner should have installed a gutter!

QUESTIONS AND ANSWERS

Question:

When my children romp around on the first level of my house above the basement laundry, dust falls out of the fiberglass insulation in the ceiling below. Is this something to worry about?

Answer:

This could be a problem if your family has allergies and the dust is moldy or contains dander from a previous owner's pet. In addition, the fiberglass could be infested with mice or even with mold-eating mites. These are some of the reasons why I don't like to see exposed fiberglass insulation in basement ceilings.

Question:

I have three small children, and my basement is partially finished. We had water damage in the basement a while ago. The walls are in contact with the cement floor. Could there be mold behind the walls?

Answer:
There could be mold growing inside the wall cavities, which could lead to exposures, because air and spores can enter the room when active children or their toys collide with walls. I would recommend removing some of the wall materials close to the floor (do this *under containment-like conditions,* to avoid spreading mold spores and potentially moldy dust), and then checking the backside of the drywall and any insulation that may be present. If you had a substantial amount of basement water, or if you find mold in the insulation or wall materials, you should hire a professional remediator rather than risk exposure yourself. Finally, any insulation that was ever wet should be removed.

TIPS

Your basement or crawl space is part of your house

- Have an electrician install lighting so you can check periodically for mold or moisture problems in a crawl space.
- Don't allow dust to accumulate on below-grade floors or walls.
- Maintain relative humidity levels close to 50 percent. (For a crawl space, you can install a remote relative humidity sensor.)

Be vigilant about conditions in unfinished, below-grade spaces

- Dehumidify unfinished, below-grade spaces.
- Goods should be stored several inches above the floor and away from the foundation walls. Use metal or plastic shelving.
- If allowed by building codes, it's preferable (except in dry climates) to have a clean, dehumidified crawl space completely isolated from the exterior and open to the basement, rather than ventilated to the outside.
- Dirt floors must be covered, either with concrete or with a vapor barrier.
- Painting floors and walls is an important part of basement hygiene.
- Keep a floor water alarm next to your hot-water heater (see Resource Guide).

- Be certain that there are no openings to the outside large enough to allow pests to get in.
- Vent the clothes dryer to the outside, never into the attic, basement, garage, or crawl space.
- An unfinished basement should be kept as clean as the rest of your house.

A finished basement is one of the most common sources of mold problems

- Dehumidify in the summer (maintaining relative humidity no more than 50 percent), and heat consistently in the winter (to around 65°F). Any isolated finished basement rooms (closets, bedrooms, bathrooms) must be similarly maintained. Keep windows closed when dehumidifying.
- Don't install built-in wooden storage benches, as mold is likely to grow inside them.
- Avoid wall-to-wall carpeting; install ceramic tile or vinyl, with area rugs if you wish to cover a hard floor.
- If the basement furniture gets a lot of use, it's preferable to have a leather couch or a futon with a mite cover on the mattress rather than upholstered pieces that can soak up body moisture and accumulate food (from people sipping and munching while watching TV).
- Keep furniture away from walls, and use furniture with legs long enough to allow adequate air space behind and beneath.
- Heating units should be kept dust free.

Certain kinds of insulation are preferable in below-grade rooms

- Avoid exposed fiberglass ceiling insulation.
- In finished basements, use foil-coated sheet-foam insulation on the foundation behind the finished walls rather than fiberglass insulation between the wall studs.

This straw hat was in a home where the furnace humidifier malfunctioned while the family was away on a winter vacation. For almost two weeks, the house air was nearly saturated with moisture. Water dripped down the windows and some of the exterior walls, which were below the dew point. Mildew such as species of *Penicillium*, *Aspergillus*, and *Cladosporium* grew on many items, particularly those with either fats (such as treated leather) or starch (such as this hat and the edges of hardcover books).

The gills of a portabello mushroom. The top of the portabello mushroom is the white fleshy cap; the gills that hang down underneath are completely covered with microscopic spores, which are released into the air. We eat the gills and spores in salads and gravies: a fungal delicacy!

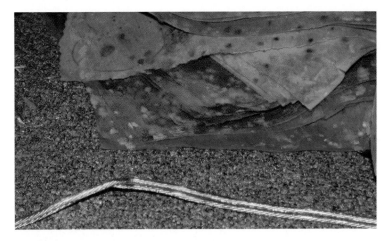

In the foreground is copper speaker wire resting on carpet in front of a basement closet. If you look carefully at the red wrapping paper, you may note something amiss with the design. Most of the "dots" are individual colonies of microfungi. The carpeting rested on concrete, and the basement playroom was not dehumidified. It is likely that microfungi were also growing (invisibly) in the dust settled within the carpet and that mold-eating mites were foraging on the mold while predator mites chased after their smaller kin. Children played on the carpet just in front of the closet.

Stachybotrys mold on the backside of drywall in a basement that flooded. A finished basement flooded with water, and fiberglass insulation remained damp for weeks before it was removed and the extensive mold problem was discovered. *Stachybotrys* mold grew on the backside of the drywall, but on the front (painted) side there was no indication of any fungal growth. Note the stains on the vertical stud at the center, indicating capillary flow of water.

bottom opposite:
Concealed mold. There was a musty odor in this room, so the homeowner pulled the paneling away from the drywall in his just-purchased house to investigate. At the left, you can see that the back of the paneling was covered with white mold, fuzziest at the bottom. At the right, black *Chaetomium* mold was growing on the back corner of the drywall paper where it touched the floor. This dark mold most often grows in soaked cellulose. This room was only one of many areas where there was concealed mold growth caused by roof leaks.

Stains showing multiple events of basement floor water. The parallel stains and the mold on the drywall indicate multiple occurrences of basement floor water. The lines do *not* indicate the level of the water after each occurrence. Instead, they indicate the levels to which water rose by capillary action from the puddle that was on the floor.

bottom opposite:

Filthy air-conditioning coil. This cooling coil from a central AC was almost completely covered with dust due to inadequate filtration. (It is very common to find a cooling coil in this condition.) The only place where clean fins are visible is along a horizontal band at the bottom. The clogged coil is very inefficient and can even lead to compressor failure. When the AC operates, moisture from the air condenses on the coil and soaks into the mat of dust, fostering microbial growth.

The shiny metal fins in the upper right third of the photo are part of the air-conditioning coil in an electric furnace. The soiled material at the bottom and on the wall of the cabinet is fiberglass that serves as both sound and heat insulation. The fiberglass lining is filthy because of inadequate filtration and poor maintenance (though the coil had probably been cleaned recently). The presence of high relative humidity led to infiltration of the dust by growing microfungi.

Aspergillus mold on the front door. A couple and two children with asthma were living in a split-level home that was kept too cold in winter. Nearly colorless *Aspergillus* mold grew on the lower, and cooler, half of the door (the flash of the camera revealed the growth). Every time someone left or entered the home, spores and micro-particles were aerosolized. A wood or metal door with a small amount of such growth can be cleaned as if it were a piece of furniture, either indoors under containment-like conditions or outdoors.

As installed on a wall, these vertical wood strips had been in contact with the concrete floor in a very wet basement. Years of being soaked by floor water led to decay of the wood by macrofungi. There are no mushrooms present and I do not believe they were picked; growing hyphae from macrofungi do not necessarily produce fruiting bodies (mushrooms or toadstools) unless the conditions are right, and the colony may have died before it matured.

bottom opposite:
Gourmet clapboards. These two beautiful mushrooms appeared at the side of a house between the time a buyer's offer was accepted and the day the home inspection occurred. There was a roof leak above the wall, and water was running down the sheathing behind the clapboards, fueling the growth of a macrofungus. As is often the case, all of the fungal mycelium and decayed sheathing were concealed by the cedar clapboards, which are resistant to fungal decay.

This cabinet was in a "finished" basement room behind a bar from which no one had served a drink for some time. On the floor above the shelves was a bathroom that had probably leaked for years. The wood is stained by water flows and discolored by micro-fungi. There were also hand prints delineated by mildew on the glass of the mirrored walls. (The "food" for the mildew consisted of skin scales, fats, and amino acids left by someone's hand pressed against the glass.)

Adequately dehumidify a below-grade space

- If you have separate rooms in a finished basement, keep connecting doors open or heat and dehumidify each space individually.
- Keep the dehumidifier free of dust and the coil clean.
- If you are purchasing a new dehumidifier, buy one that has a filter and consider using a larger-capacity machine, such as a Therma-Stor Santa Fe® (see Resource Guide).
- In humid climates, a "dehumidifier" that merely exhausts air from a basement is not useful. Use a machine that dehumidifies by condensing moisture from the air.

Minimize sources of moisture

- Grading at the exterior should be sloped away from the foundation.
- Keep your gutter system clean to avoid clogging and overflowing, and use downspout extensions to prevent ponding at the foundation.
- If you don't have gutters, be sure there is an adequate roof overhang.
- Basement windows and doors, including the bulkhead and the door leading from the basement to the upstairs rooms, should be closed when dehumidifying, to prevent the entry of moist air.
- A major flood, as well as any sewer backup, should be dried up and cleaned as quickly as possible by trained professionals under containment conditions (see part 3).

Chapter 6

MOLD IN THE MECHANICALS

Any mechanical equipment that supplies warm, cool, dehumidi-fied, or humidified air is prone to mold growth and can thus be a major source of airborne fungal allergens. With proper maintenance and the best possible filtration, however, you can go a long way toward controlling the release of bioaerosols into indoor air.

Hot-Air Heat and Central Air Conditioning

In the Northeast, central air conditioning (AC) is relatively rare in older homes, unless owners have added AC to an existing hot-air heating system or as a separate system altogether. In New England and in other parts of the country, most new homes are built with either a combined hot-air furnace / AC system or a heat pump that supplies both heating and cooling. Sometimes a system is divided into two or three zones so heating and cooling can be independently controlled in different areas of the house.

Hot-Air Heat

In a hot-air furnace, metal resting on the concrete floor can be below the dew point of the summer air, and mold can grow inside, in the dust on the bottom and walls of the blower cabinet. This is why I always recommend that systems be properly supported an inch or two above the floor or be insulated from the concrete.

Ducts that pass through unconditioned (neither heated nor cooled) space, such as a crawl space, can also harbor mold within them. In the summer, house air may slowly flow through the ducts by convection, even though the heating system is turned off. That air may be warmer than the air in a cool crawl space (or basement). Air within the ducts is cooled by heat loss to the metal, causing higher relative humidity and the potential for mold growth in the dust. Insulating the ductwork can help solve this problem, but in the end it is preferable to avoid installing ducts in uninsulated crawl spaces.

Central Humidification

Some people find that the air in a home with a hot-air heating system feels dry, and they install a furnace humidifier. Most furnace humidifiers contain a water reservoir, and inevitably, dust accumulates and microbes (mold, actinomycetes, and bacteria) proliferate. Some humidifiers have a drum wrapped with a sponge that rotates in the wretched broth, dispensing microbes and the products of their growth into the supply system. For this reason, and because I have seen hundreds of such humidifiers broken, leaking, and abandoned, I don't recommend any type of centralized humidification system that has a water reservoir.

In a spray humidifier, water is sprayed as a fine mist into the cool (return) or hot (supply-side) air of a furnace. I don't recommend this type of humidification either, and its use has more or less been abandoned in residential systems because of the risks associated with spraying water into the ducts. (Not all the water evaporates, and the dust within the ducts can get wet and moldy.)

A third kind of system has a metal mesh pad with water trickling over it and hot air blowing across the mesh. The excess water is pumped out of the system (requiring a condensate pump). This type of humidifier, if maintained, should not develop microbial problems.

Whatever type of furnace humidifier is installed, it should be checked weekly for leaks and cleanliness. Refer to the manufacturer's directions for proper cleaning and maintenance.

Central Air Conditioning

In an AC system (as well as in a hot-air system), the temperature is controlled by a thermostat. Despite its gradations in temperature settings, the system is either on or off. Unfortunately, a thermostat

Leaking furnace humidifier. The white plastic water reservoir is at the bottom of the furnace humidifier. Below a horizontal seam in the humidifier are stains running down the edges and center of the vertical return duct. The stains consist of oxidized zinc, rust, and minerals from the water that overflowed from the humidifier reservoir. The water also leaked *into* the return duct, where it provided moisture for the mold and mites flourishing in the dust.

in an AC system directly controls only the temperature, not the relative humidity (RH), and thus in my opinion is inadequate. I believe that any cooling system should be designed to control temperature and RH separately (temperature with a thermostat and RH with a "dehumidistat"). If you have a dehumidistat, you can maintain the RH at 60 percent and keep the thermostat at 72°F when you are at home but let the temperature rise to 80°F when you are away. The dehumidistat will still activate the system to remove moisture when needed. Controlling the relative humidity indoors in this way will allow you to prevent many, though not all, of the problems that occur as a result of condensation indoors. Unfortunately, few manufacturers now provide AC systems with such dual control, so one way to accomplish dehumidification and cooling separately is to install a central dehumidifier (such as a Therma-Stor®) that is independent of the central AC controls but uses the same duct system.

In an AC system, mold can grow in the dust on the blower or wet cooling coil, as well as on both sides of the supply registers. This makes sense when you remember that cold air (sometimes as cool as 55°F) exits the metal register (or grille) of a supply duct. When the blower turns off, the metal register may end up below the dew point of the room air, leading to condensation and growth of microfungi in the dust. Sometimes colonies are visible as black dots on the metal, but often dust full of microfungi looks no different from ordinary dust—another reason to keep registers visibly clean.

If excess relative humidity exists in the ducts, or if the fiberglass lining material has been soaked by leakage from a clogged condensate pan (see "Leaks" later in this chapter), microfungi will proliferate within, because the fiberglass is often full of nutrient dust as a result of inadequate filtration. Air-conditioning ducts buried in concrete are also prone to mold growth, because the interior surface of the ductwork is cooled by heat loss to the concrete.

Some kitchen cabinets or bathroom vanities have a vertical AC register in the kickplate just above the floor. Often the exiting cold air will fill the gap (being used as a "duct") between the bottom of the cabinet and the floor. This air cools the wood floor, and if the cabinet is located over a humid crawl space, the underside of the floor (particularly in Southern climates) will be digested by fungi. (The simplest way to avoid such problems is to be sure the soil is covered with a vapor barrier and to dehumidify rather than ventilate the crawl space.)

In the Northeast I commonly see an older home with a hot-water boiler for heating and a separate AC system located in the unheated attic. Very often the large opening for the filter and return grille is installed in the ceiling of a hallway, just outside a bathroom. In the winter, warm moist air from showering rises by convection into the cold duct system and condenses on the dust-covered duct walls, leading to extensive growth of *Cladosporium, Alternaria,* or *Aspergillus* microfungi in the dust. When the system is turned on in the summer, spores or particulates contaminated with microbial allergens may be dispersed into the airflows. To prevent heat loss and condensation in such a system in the winter, seal the return at the grille (by covering the filter with foil) and close the supply registers; do not forget to open everything up again in the spring before turning the system on.

It's also common to see mildew (microfungi) growing on surfaces that are located beneath uninsulated or poorly insulated AC supply ducts, because water vapor can condense from the air on the cooler surfaces of the ducts and drip below. Some people who buy an older home with a hot-air furnace modify the system to include air conditioning. Unfortunately, in most of these homes the older metal ducts are not insulated, and as cold air is circulated through the system, condensation may occur on the outside of duct surfaces if they are below the dew point of the house air. In the basement of one such house, *Stachybotrys* mold was growing in the ceiling tiles under the dripping ducts. If you want to add central AC to an older home, insulated ducts should be installed.

In humid spaces, the metal boots where the ducts terminate should also be insulated. In damp basements and crawl spaces, I even see *Cladosporium* microfungi covering the vinyl wrap of fiberglass duct insulation, because the surface of the insulation is below

Filthy air-conditioning coil. This cooling coil from a central AC was almost completely covered with dust due to inadequate filtration. (It is very common to find a cooling coil in this condition.) The only place where clean fins are visible is along a horizontal band at the bottom. The clogged coil is very inefficient and can even lead to compressor failure. When the AC operates, moisture from the air condenses on the coil and soaks into the mat of dust, fostering microbial growth.

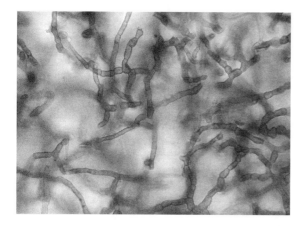

Mold hyphae growing through dust on an AC coil. These hyphae are growing in the mat of dust on the coil in the previous figure. The debris in the mat is held together by the hyphae twining between the dust particles. A mat like this usually contains a variety of microorganisms, including mold, yeast, and bacteria. (400x light)

Stains and corrosion from condensation on an uninsulated duct. The bottoms of two ducts—a supply *(left)* and a return *(right)*—and stained framing are visible in this basement closet, where the moldy ceiling had been removed and a roasting pan (not visible) had been set on the top shelf. The chase for the ducts went from the basement all the way up to the attic, where the top of the chase was open, allowing humid outside air to condense on the cold supply duct and drip down onto the wood and ceiling.

the dew point of the air. Lowering the relative humidity of the basement or crawl space is one solution to this problem. Again, though, it's best if possible to avoid having AC as well as hot-air ducts pass through below-grade, unconditioned crawl spaces.

Since there is cold water in an AC condensate line, if the line runs inside a wall or through any humid space in a damp climate, water can condense on the outside of the pipe and cause mold growth. To stop condensation, the pipe and condensate trap should be wrapped with foam insulation. In fact, check to see that any cold pipe leading to AC equipment is well insulated (and be sure there are no gaps in the insulation).

Cooling in Dry Climates with an Evaporative Cooler

In climates where the air is arid, an inexpensive alternative to air conditioning is an evaporative cooler (sometimes called a *swamp cooler*). Such a unit sits on a roof or at the side of a house and consists of a blower to draw in and circulate air, several porous pads, a water reservoir (with a float valve), and a pump to circulate water. The pump takes water from the reservoir and wets the pads as the blower pulls outside air across them. Evaporation of water from the pads cools (and humidifies) the air, which is then circulated through a duct system to the interior. (In Arizona, for example, dry air at 90°F might be thus cooled to 75°F.)

In communities where the outdoor air is clean, properly maintained evaporative coolers should not experience mold problems. In places where the air contains large concentrations of spores, pollen, and other plant materials, however, the potential exists for the accumulation of biodegradable materials on the pads (which act as filters), and hence the chances of mold growth are greater, particularly since the accumulated debris is kept wet. In addition, if the supply duct located near the evaporative cooler is lined with fiberglass insulation, dust in the insulation can become damp and moldy.

Maintenance should follow the manufacturer's instructions and may include regular treatment of the water with antimicrobial additives (and/or replacement of the pad as needed), servicing of the float valve (which can become clogged with minerals), cleaning of the blower when soiled, and draining and sealing at the end of the cooling season. Since these units do not have returns, they pressurize the house with air (see chapter 2), so for such a system to be effective, windows must be kept open very slightly (about an inch) in the rooms where cooling is needed, to allow the entry of cooled air and the flow of excess air out the windows.

Some homes have ceiling relief grilles that serve as vents and allow the excess air to flow into the attic; in that case, windows do not have to be opened. Since these vents are leaky, they should be sealed during the winter to prevent heat loss to the attic. During rain or cool and humid weather (or if the relative humidity indoors rises above 65 percent), the evaporative coolers should not be operated, in order to avoid the risk of mildew.

Leaks

Central AC systems may develop leaks. One family contacted me because two days after they moved into their new home, water dribbled down a wall. They found that the attic AC condensate line was leaking from a loose joint. The problem was repaired, but they were still worried that mold might be growing inside the wall, and they didn't know what to do.

I recommended that they first talk to the technician who had repaired the condensate line, to get an opinion on how long the leak had lasted. If the leak first occurred when they turned the AC system on and the wall was not insulated, then most probably any damage was only cosmetic. On the other hand, if the wall contained insulation, or if the AC had been operating and leaking for weeks during construction, it was possible that there could be concealed mold.

The most common source of AC leaks is the condensate pan overflowing after the drain has clogged with "bioslime" (due to inadequate filtration). If you've had such a leak and you are worried, I suggest you send a sample of the dust from the fiberglass liner to a lab to see if there is microbial growth, but if the liner has been soaked, it should be replaced by a professional. (Remember to wear a NIOSH N95 mask whenever disturbing potentially moldy dust.) To minimize cosmetic damage from such leaking, I recommend that you have a technician install an overflow tray under any attic unit. At the same time, install a shut-down float switch in the tray; that way, if the AC ever overflows, the water in the tray will activate the switch and turn the system off.

Duct Cleaning

Ducts can be made out of metal, rigid fiberglass board, or flexible, fiberglass-insulated plastic. Metal ducts are easiest to clean, because they have a smooth, hard interior surface. Fiberglass ("duct board") ducts have a coarse, porous surface and can be damaged by cleaning (though I hear that newer products have a coating that makes the surface less porous). Flexible ducts have interior ridges (due to the metal spiral coil installed to stiffen them) that make them difficult to clean. Whenever feasible, it's preferable to replace rather than to attempt to clean mold-contaminated fiberglass or flexible ducts.

I read about one poor woman in Texas who hoped to have her ducts cleaned for the advertised special rate of $99. The crew arrived and then told her that her entire duct system was full of mold. They recommended an expensive cleanup procedure, including an antibacterial treatment of the ducts, at a cost of over $500. (This bait-and-switch technique is apparently not uncommon.) In the weeks that followed, her duct system developed a musty smell. Her insurance agent inspected the system and told her that the ducts were made of fiberglass duct board, not metal. These ducts have aluminum foil on the outside, but their exposed internal surfaces are made of porous fiberglass, and they should never be treated with chemicals and brushes; in fact, residential fiberglass ducts like these, when contaminated with microbial growth, have to be replaced (though in large commercial systems they are sometimes spray-coated with a sealant). In this case the entire duct system was ruined, and it cost the homeowner over $3,500 to replace it. A bit more than the $99 special!

"How often should I have my metal ducts cleaned?" is a question I hear frequently. Ductwork is only part of an air conveyance system, which also consists of the blower, the blower cabinet and the plenum, and, in a heat pump or central AC system, the cooling coil as well. All of these components may need cleaning, not just the ducts.

The Environmental Protection Agency (EPA) has published a booklet called "Should You Have the Air Ducts in Your Home Cleaned?" (see Resource Guide). The booklet states that "duct cleaning has never been shown to actually prevent health problems." This statement is technically correct, because having your ducts cleaned is no guarantee that you will not become sick. On the other hand, the statement could be misconstrued to mean that duct cleaning has never been shown to improve anyone's health. In my experience, mold-contaminated air conveyance systems are very common, and are often the cause, at least in part, of many occupants' respiratory problems. In most cases of "sick" houses, after a contaminated air conveyance system has been properly and professionally cleaned (and filtration improved), the sufferers experience a reduction in their symptoms.

In new construction, it is common for the contractor to operate the heating or air-conditioning system during dusty operations. As a result, sawdust and drywall dust can accumulate in the air conveyance system. (For families with severe allergies, I recommend that, if possible, the entire duct system be sealed immediately after installation, and that, if necessary, some form of alternative heating or cooling be used during construction.) Before people move into a newly constructed or renovated house, ducts should be inspected (with a mirror and flashlight) and cleaned if construction dust and debris are present. If a musty odor is coming from your hot-air heating and/or central air-conditioning system, or if a sweat-sock smell is coming from your central AC, there is likely to be microbial growth within, and you need to hire a professional duct cleaner. Otherwise, in my opinion, ducts don't really need to be cleaned more than once every five or ten years. Very contaminated lining materials should always be replaced, using great caution to avoid spreading mold spores indoors. In addition, where possible, the new insulation installed in the vicinity of the AC coils and blower should not have a porous surface. (Closed-cell foam approved for the purpose is one acceptable type of insulation, and foil-covered fiberglass is another.)

There are different methods for cleaning ducts, but I only recommend the type of cleaning that makes use of brushes to wipe the dust off the insides, and a HEPA (high-efficiency particulate arrestance) or truck-mounted vacuum to clear the debris. During the cleaning process, it is very important in the homes of sensitized individuals to prevent the release of potentially allergenic particles into the house, so be sure to ask duct cleaners how they plan to contain the dust. I also recommend that you obtain a copy of the EPA booklet mentioned earlier, and only use a company that belongs to the National Air Duct Cleaners Association (NADCA).

Finally, be sure the duct cleaners properly seal any openings they have made. In one home where the ducts had just been cleaned, hot air was pouring out of a supply and moldy crawl space air was being sucked into a return where openings had not been properly resealed using a patch larger than the opening, fastened by screws and tape. In fact, when the duct cleaners arrive, ask them to check that the visible joints in your duct system are airtight. If necessary, hire a mechanical (heating and cooling) contracting company to check this or to make repairs.

Some duct cleaners recommend that a "bacteriacide" be sprayed into a duct system after cleaning, but I do not believe in the routine use of any type of chemical in ductwork. Though approved sanitizers may have to be used on the AC coil, very few chemicals have been approved by the EPA for such use. As discussed earlier, some of the less reputable duct-cleaning companies entice homeowners with very low prices and then talk them into buying an expensive, unnecessary spray treatment as well. In my opinion this is an unethical business practice, and the cost of this treatment may be several times greater than the cost of the cleaning. Moreover, mold can still grow in the dust that begins to accumulate soon after treatment.

Filtration

Ducts can be new and sparkling clean, but if the AC unit or furnace is contaminated with moldy dust, the system may still disseminate spores or allergens. Protecting the cleanliness of the system with proper filtration is therefore essential. Some people think that the purpose of a filter in an AC or hot-air heating system is to "clean" the house air, but a filter's primary job is to keep the blower cabinet, the blower, and the AC coil (if present) free of dust. Even the most efficient filter in the world will not prevent the release of contaminants, however, if any one of these components is coated with dust and microbial growth. Thus, filtration and cleanliness are partners in maintaining the system's hygiene.

Most of the soiling in a residential duct system takes place on the return side (return grille, ducts, and blower cabinet), because this is where most of the particulates (human skin scales, clothing lint, pet dander, etc.) are being drawn into the system. In humid locations, very soiled returns can become moldy and support the growth of mite colonies. One way to keep the return ducts clean is to install a holder for a coarse fiberglass prefilter (*inefficient,* so the airflow restriction is minimal) and change the filter annually (on smaller return grilles, coarse filter fabric can be cut to size and fastened behind the grille).

You must nonetheless have a *pleated media filter* just in front of the blower and coil to prevent the accumulation of nutrient dust on the blower and coil surfaces. There are many different kinds of filters available and, unfortunately, at least four different systems for rating them. For example, the typical fiberglass furnace filter has approximately a 75 percent *arrestance,* a 15 percent *dust spot efficiency,* and close to a zero percent *DOP efficiency* (DOP, or dioctylphthalate, is a chemical mist used to test filters). In other words, the same filter could be described as having about a zero percent up to a 75 percent efficiency, depending on the type of testing. And the efficiency also depends on the airflow rate through the filter, as well as on how old it is. The dirtier the filter, the more efficient it becomes, but a dirty filter can also restrict airflow, which is why it has to be replaced.

The newest rating system is called *MERV (minimum efficiency reporting value)* and is probably the most reliable way to rate filters, but it's not linear. In other words, a filter with a MERV rating of 8 is not just twice as efficient as one with a rating of 4, but is many times as efficient. Ideally you don't want a filter with a MERV rating less than 6. A one- or two-inch pleated media filter with a MERV rating of 8 captures about 70 percent of smaller mold spores (those measuring around 3 microns in diameter) and a greater percentage of the larger mold spores, as well as pollen (most of which are larger than mold spores). Keep in mind, though, that the higher the efficiency, the greater the resistance to airflow. Because a deeper media filter has more pleats, it may last longer than a one-inch filter. Before installing any such efficient filter, check with the manufacturer and the system's installer on the compatibility of the filter with the equipment.

Make sure any filter is properly sized and sealed in its enclosure. The advantage of a one-inch media filter is that it can be inserted to replace your existing filter, but with minor modification by a technician to the filter holder you may be able to fit a two-inch media filter. Four- and six-inch media filters require duct system modification and can only be installed in the manufacturer's housing.

Good filtration in any system is essential to keep the blower and coils clean, and the MERV 6 or 8 filter should only be installed close to the blower, rather than far away at a remote return grille. Always change the filter according to the manufacturer's directions, but at least once or twice a year. (A clogged filter can reduce airflow so much that it causes an AC or heat pump compressor to fail!) Some more efficient filters may have to be changed more often.

Never try to clean a disposable filter with a vacuum cleaner or to reuse it by turning it around; just throw it out and replace it. Don't use a thick washable filter, because you can never clean it well enough or wash it often enough, and if the filter stays damp and dirty, mold can grow in the dust. (Thin metal or plastic mesh filters can be washed, but they are too inefficient, with a MERV 3 rating, to be used alone.) I also don't recommend electronic filters, because they are very expensive, and homeowners rarely maintain them properly. (In my opinion, electronic filters should be cleaned at least monthly.)

How can you tell if the filter you are using is inefficient? If you stand in front of a mirror and hold a new, clean furnace filter up to your face and you can still see your reflection, then the filter will not trap enough of the dust.

In homes, supply ducts usually have less accumulated dust than return ducts have, because most of the particulates have been filtered out, particularly if a pleated media filter is in place. In a hot-air heating system, supply ducts in wall or ceiling cavities generally do not get moldy, even if dusty, because hot air passes through them, creating dry conditions that are not conducive to microbial growth. (But remember, dust in furnace supply ducts in crawl spaces or basements, or anywhere in an AC system, *can* get moldy because of high relative humidity or near dew point conditions. A kitchen floor duct can also get moldy if food falls in, particularly if near an exterior wall. And never put wet boots on a floor grille, because you will be watering the dust below.)

Dehumidifying Above-Grade, Habitable Spaces in Damp Climates

In some locales the weather is humid but not very hot. In these areas, central or portable dehumidification of air in above-grade spaces may be a viable, cost-effective alternative to cooling with central air conditioning. Vacation homes at the seashore are particularly prone to the proliferation of mildew, and during periods of vacancy, when cooling is not needed, dehumidification can be a beneficial, low-cost alternative to removing moisture with air conditioning. Since the mechanical process of dehumidification involves the generation of heat (by a compressor), central dehumidification of habitable spaces usually requires a small, separate source of air conditioning to cool the house air.

Other Kinds of Heating Systems

Of course, a hot-air heating system is not the only option. Some people have electric heat. One family lived in a slab-on-grade house, and they could smell mold inside and outside the building. They ran a dehumidifier all year round, but it didn't seem to help the situation. When they closed the house up tight in the winter, the smell got worse.

The fact that they had to run a dehumidifier in all seasons suggested to me that they were generating too much moisture in the house and creating dew point conditions. First, I encouraged them to be sure that they had adequate exhausts for cooking and showering, and that the clothes dryer was vented to the outside. In addition, I suspected that they were turning the thermostat down too low during the day. Electric heat is much more expensive now than it was years ago, so people often turn the heat way down when they leave the house for work. This creates high humidity at cold exterior walls, often leading to mold growth. (In one electrically heated home, mold grew visibly on walls in vertical stripes at the studs, where the walls were colder. See the discussion about soot patterns in chapter 7.) I always advise people with electric heat to keep the indoor temperature at 64°F or higher, even when no one is home. (Think of the extra heating expense as the cost of a health benefit.) Finally, I suggested that the family have dust from the carpeting tested for microbial growth and, if necessary, replace the carpeting or install wood or vinyl flooring.

Mold can grow in the settled dust on any type of heat emitter, including radiators and baseboard convectors in steam or hot-water systems. For example, I have found extensive growth of *Penicillium, Aspergillus,* and *Cladosporium* microfungi in the accumulated dust on the undersides of baseboard convectors in forced hot-water systems, particularly in rooms partially or entirely below grade. (Check the bottoms of heat emitters with a mirror and flashlight.)

In very old homes I have occasionally found the dust between the sections of a radiator to be full of mold and other allergens. It is therefore essential that the outside of radiators and the walls behind them, as well as the undersides of baseboard convectors, be HEPA-vacuumed of all dust prior to moving into a house. (It's also not a bad idea to HEPA-vacuum all heating units—radiators, baseboard convectors, heat registers—at the start of every heating season.)

One man called me because he was sensitive to mold and was worried about living in a house with steam heat. The concern, of course, is to prevent moisture from entering the house air. In a steam heating system that is functioning properly, very little steam or water should exit the system. On the other hand, in homes where the radiators aren't maintained, a steam valve or an air vent may leak water or steam. To prevent mold from growing on the wall and floor nearby, a leaking valve should be fixed and a malfunctioning air vent replaced. It's also important to check your boiler periodically for leaks.

Sometimes in homes with hot-water or steam heating systems, a tankless coil is attached to the boiler to heat the "domestic" water (the water used for showering and washing). To operate this hot-water system, the boiler stays on all year round, even in the milder months. In consequence, the basement is warmer, and therefore the relative humidity is lower. I usually find far less mold growth in basements with this kind of system.

One woman was considering installing radiant floor heating in her basement ceiling, to keep the below-grade family room as well as the kitchen above drier and warmer. I think radiant heat is a great

Leaking air vent on a steam radiator. This vent should have been replaced years ago, because it leaked steam that condensed on the wall. Water dripped onto the floor beneath, and mold grew in the dust on the cold exterior wall and the floor below.

idea, but the ideal answer for a comfortable, finished basement space, if you can afford it and have the necessary ceiling height, is radiant heat in the basement floor itself.

Portable Equipment

Portable equipment such as a window air conditioner or a humidifier usually contains moisture and dust—the perfect mix for microbial growth.

Window or Wall Air-Conditioning Units

Unfortunately, none of the common AC window units that I have seen are supplied with adequate filtration (some manufacturers' filters have a MERV value of only 2), so dust accumulates on the wet cooling coil, and mold (as well as yeast and bacteria) will grow. Often, when you look beyond the supply grille (where the cold air comes out), you can see black dots on the interior walls; these may be *Cladosporium* mold colonies. If the filter is touching the damp coil, it too can become contaminated with microfungi. Finally, because of the dew point condition of the air, any dust on the blower or walls of the discharge side can also be consumed by mold.

If a window AC has barely adequate filtration and you leave the unit in place all year round, year after year, spores and other allergens may be dispersed into the air when the unit is turned on in the summer. For most people this will not be a problem, but in a family with significant mold allergies, I recommend that, where feasible, a portable AC be removed from its case at the start of every cooling season and be taken outside to be cleaned and disinfected with a dilute bleach solution, taking care not to wet the electrical components. (Servicing of portable units should probably be left to professionals.) If it is too difficult to remove a portable AC from its case for cleaning, be sure that you use efficient filtration and start each cooling season with a new filter in place.

When the AC is reinstalled after cleaning, or when a new AC is installed for the first time, always make sure the condensate leaks to the outside, and never to the inside, to avoid causing microbial growth on the wall or in the carpeting beneath. This is accomplished by having the unit tipped (according to the manufacturer's directions) so that the outside end is lower than the inside end.

In my opinion, no type of AC equipment, whether portable or central, should be operated with a filter rated less than MERV 6. I recommend that people use a pleated media filter on central systems and a high-efficiency filter fabric (cut to proper size) on a portable AC. (Check with the manufacturer on the highest allowed MERV rating, but be sure the filter does not touch the cooling coil.)

Humidifiers

Some people run portable humidifiers for their own comfort, or for the sake of plants or musical instruments, and as a result can potentially create high relative humidity where indoor air is near colder exterior surfaces. Portable humidifiers can also develop mold problems within. I once took a sample of a black stain from a piece of cloth draped over a heating bar that was part of a humidifier system for the inside of a grand piano. (Apparently pianos stay in tune if maintained at 42 percent relative humidity.) The cloth, which was made of cotton (cellulose), was covered with *Stachybotrys* mold. I don't think the contaminated cloth was producing any airborne spores, because it was not disturbed by anyone while it was wet and inside the piano, but if the homeowner had removed the cloth when dry, this very cultured colony could have produced airborne spores.

There are three different kinds of humidifiers. "Ultrasonic" humidifiers create a visible cloud of water droplets by vibrating a small metal plate immersed in water. "Cool-mist" (or evaporative-pad) humidifiers create water vapor by blowing air over a wet pad to evaporate water at room temperature. "Warm-mist" humidifiers produce water vapor and steam by boiling water. I don't usually recommend the first two types of humidifiers. Ultrasonic humidifiers can aerosolize dissolved minerals and contaminants that are present in the water. In several evaporative-pad humidifiers I have found *Stachybotrys* mold growing on the immersed part of the pad. I think the only humidifier safe from microbial growth is the one that boils water, the warm-mist humidifier, because the conditions are too hot for growth in the dust suspended in the liquid. If you choose a warm-mist humidifier, be sure to get one that has a humidistat, and follow the manufacturer's cleaning suggestions.

QUESTIONS AND ANSWERS

Question:

What's the deal with humidifiers? Some say *never* use them because they grow mold and bacteria. Others say if the relative humidity falls below 30 percent, sinus membranes will dry out. What do you recommend?

Answer:

Relative humidity under 30 percent in the winter can be uncomfortable for many people. The danger, though, is that if you overhumidify, condensation will occur on cold walls, leading to mold growth. That's why I recommend that, when possible, the relative humidity should be maintained slightly above 30 percent in the winter, that any humidifier be equipped with a humidistat, and that the air be monitored with a thermo-hygrometer. In very cold weather, check for condensation on the inside surfaces of windows. If the surface of the glass is below the dew point of the indoor air and excessive moisture condenses, you will have to lower the relative humidity (or if you cannot, install storm windows or replacement windows with insulated glass).

TIPS

Hot-air heating and central AC systems can be major sources of mold contamination

- All blower cabinets should be free of dust, moisture, condensation, and water leaks.
- Blower cabinets, filter holders, and ducts should be airtight.
- Ductwork should never pass through concrete in contact with soil.

- Any ductwork that passes through a completely unconditioned space (attic or crawl space) should be well insulated. In new construction, if you have the choice, run ducts through conditioned spaces.
- If you have central AC, try to keep the relative humidity under 60 to 65 percent in the cooling season. If necessary, use supplemental dehumidification. (If you have a furnace humidifier with central AC, be sure to empty, clean, and shut down the humidifier for the AC season, or you will be humidifying your summer air.)

Pay attention to filtration, central humidification, and duct cleaning

- Assuming adequate filtration is in place, ducts need to be cleaned no more than every five to ten years, or inspected and cleaned if construction dust and debris are present.
- Use a pleated media filter with a MERV rating of at least 6, but preferably 8. (Check with the manufacturer and/or installer about the compatibility of any filter with your system.)
- Avoid having bacteriacide or fungicide sprayed into the ducts.
- If you insist on having a central humidification system, use the flow-through type (no water reservoir) with a metal mesh pad and a condensate pump.

Other heating systems can also be associated with mold problems

- If you have electric heat, keep the indoor temperature at 64°F or higher during the heating season (winter in the United States), whether you are home or not.
- Keep radiators and baseboard convectors as free of dust as possible.

Fungi can grow in portable AC units and in humidifiers

- In families with significant mold allergies, portable AC units should be professionally cleaned and disinfected, preferably at the start of every cooling season. Use MERV 6 filtration, but check with the manufacturer on compatible filter types.

- Humidifiers should be kept free of dust.
- I do not recommend that people use humidifiers with evaporative cellulose pads.
- To control the relative humidity, choose a warm-mist humidifier with a humidistat.

Chapter 7
THE SPACES WE LIVE IN

Microfungal growth is often present where we least expect it, sometimes invisible, sometimes right in front of us: on an inherited grandfather clock outside a master bedroom, on the bottom of a drawer in a dresser picked up at a yard sale, in a used TV that was stored in someone's basement, or inside a pillow in a vacation home. In the rooms we inhabit, there are many unsuspected locations for mold growth.

Furniture

As discussed earlier in the book, when the relative humidity (RH) is over 75 percent, microfungi (most often of the genus *Aspergillus*) can grow unseen on the unfinished wooden undersides of couches and chairs, the lower shelves of tables, and the bottom drawers of bureaus that have been placed in below-grade spaces or stored in unheated, damp garages. Wooden furniture and shelving placed up against cooler exterior or foundation walls can also house mold growth. High RH can also lead to mold growth in the cushions of upholstered furniture, which may contain mold spores settled from the air and all sorts of food for the microfungi (skin scales, cookie crumbs, pollen, and pet dander). People can even provide some of the moisture needed for mold to grow in upholstered furniture; a

Bedpost with *Aspergillus* mold. There is a child's scrawl at the top of the bed frame at the right, but the vertical bedpost is covered with mold. The bed was in a below-grade bedroom, and the child who slept there suffered allergies.

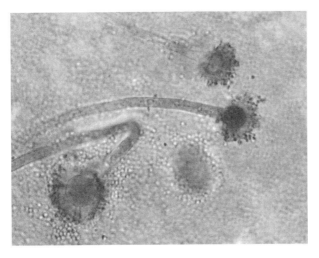

Aspergillus mold from a bedpost. Four *Aspergillus* conidiophores along with hundreds of individual spores were present in the sample of the dust from the bedpost in the previous figure. (400x light)

teenager who spends hours slouched on a couch like a setting hen is dampening and possibly incubating the fungal (and mite) nest.

One woman experienced mold allergy symptoms whenever she returned to New England to visit her mother for the holidays. The mother lived alone in a large house, and she turned the heat up in the living room only when her children came to visit. As a result, the air in the nearly unused room, located above a crawl space, was often cold and damp, humid enough near the floor for mold to proliferate. Not only were the chair legs covered with microfungi, but when I looked at a sticky-tape sample of the growth, I found live mold-eating mites. In another family the young daughter began suffering asthma symptoms as soon as she entered her home, despite her parents' dedicated cleaning and vigilant attention to home environmental details. In the living room, an easy chair contained mold, and a couch contained mites. They temporarily sealed these pieces using plastic sheeting and duct tape, and the child's health improved so much that the parents were able to stop administering her nightly asthma treatments.

Walls, Ceilings, and Floors

In cold climates, mold can grow on the habitable side of inadequately insulated exterior walls, particularly where the conditions are coolest: close to the floor (because cold air sinks) and in outside corners. Unheated closets, where one or both walls face the outside, can also be a problem. Clothing stored in moldy closets can get contaminated with spores from the dust, and if the clothing is in contact with the cold wall, mold can begin to grow in the fabric itself. If you have a closet where mold grows, you can buy a small warmer, but be sure it's safe and was manufactured for such use (clothes should not come into contact with heating elements). You can also install a louvered door to the closet or insulate the exterior walls. If you are building a new house with corner closets that have exterior walls, be sure a component of the heating system warms the closet.

People also find mold on interior window trim, when water has condensed on the glass and dripped down to wet the wood or dust

Stachybotrys mold on a closet wall. Water from a roof leak dripped unseen into the cavity of this closet wall. The owner thought he could "repair" the wall by patching and painting, but he never fixed the roof leak. Three oval colonies of *Stachybotrys* mold grew right back through the wet paper of the patched and painted drywall.

below. One family moved into a new home and noticed mold on windows, woodwork, and walls. They kept cleaning the mold with a bleach solution, and even repainted using a paint that contained mildew inhibitors, but the microfungi continued to grow back. They had to reduce the RH in their home to prevent the reoccurrence of mold growth. I also recommended that they use a HEPA (high-efficiency particulate arrestance) vacuum on their rugs and furniture to reduce the number of spores that might have settled into the house dust as a result of their frequent cleaning of the mold. Finally, I told them that since mold usually grows indoors in the dust on painted surfaces, rather than in the paint itself, inhibitors added to the paint are not very useful, in my opinion, and some may even off-gas low levels of potentially toxic vapors. One of my clients had to sell his house after a contractor used paint with exterior mildewcide in a finished basement room; another homeowner had to gut the entire interior of a home after sealant with toxic mildewcide, intended for outside use, was used on all the interior wood panel walls.

Ceilings and walls directly beneath the roof are vulnerable to staining and fungal growth due to concealed roof leaks. I received an e-mail about a one-level house in which a roof leak had been repaired by a contractor, but the living room seemed to have new signs of water damage. The homeowner bleached the area, but the stain still seemed to be spreading, and soon the ceiling developed an unpleasant musty smell, probably caused by microbial growth. I suggested that she very carefully mark with pencil or chalk the perimeter of the stain, because once a leak is repaired, the stain should not expand. Any further spreading of the stain would indicate either that the leak had not been properly repaired, or that a new leak or soaked insulation was present.

In another home the uncarpeted wood floor and door trim near a sliding door leading to a deck was turning black. Was this mold, the homeowner wondered? Some finishes will darken when exposed to sunlight. (In addition, when a wood like oak is frequently dampened by pet urine, it may turn black because of microbial growth.) Darkness on varnished wood flooring close to a sliding door leading to a deck, and on varnished or unfinished wood at the bottom of vertical door trim, is often caused by mold growing because roof water splashes onto the deck and soaks into the slider, frame, trim, and/or flooring. The water may fill the tracks of the sliding door and leak into the interior.

If you have a deck and the flooring near the door is turning black, you can test whether rainwater is entering the interior by pouring a cup or two of water directly into the tracks and looking carefully to see where it is flowing (hopefully, to the outside rather than to the inside). Don't forget to look under carpeting, if present, and go into the basement to see if water is soaking the ceiling or the framing under the slider; if so, you may have mold problems there as well. In addition, also check the threshold and basement during a heavy rain, because wind-driven rain may wet the siding and trim around the door and leak into the wall or basement. Finally, roof water should never be allowed to drip directly onto a deck below, because the splash always causes problems; install a gutter for the length of the deck.

Rooms with Water

Ironically, the water in bathrooms and kitchens that makes these rooms function as intended is the same water that, if not properly contained, can lead to mold problems.

Bathroom Walls and Ceilings

You can often find mold on the upper part of walls and on ceilings in bathrooms, because these areas are by their nature prone to conditions of high RH (and because hot, moist air rises). In addition, in humid weather, when the temperature of the water in the toilet tank is below the dew point, moisture will condense on the tank and drip onto the floor below, which can lead to mold growth. To prevent the potential for mold growth, you have to minimize the moisture. In a bathroom this is obviously difficult, but it can be done.

After showering, run the exhaust fan, and leave the bathroom and shower door or curtain open. If you have space on the vanity top, you can even operate a small table-top fan to mix the air and hasten evaporation (but be sure the fan is plugged into an outlet that is protected by a ground-fault interrupter, or GFI). If there's a window, leave it open in mild weather. If you are really concerned (and particularly fastidious), use a squeegee on the shower walls. If it's any comfort, mold on bathroom ceilings, though quite common, is not, in my opinion, as much of an exposure problem as mold in a carpet, because the growth is not as readily disturbed. Nonetheless, there should not be visible mildew in a bathroom.

If you have mildew problems in a bathroom, consider drying damp towels in another room, because they are just one more source of moisture. Small rugs that get wet should also be hung elsewhere to dry, as well as washed periodically. Fleecy surfaces such as these can harbor mold, mites, and bacteria if left wet for too long.

A persistent mold problem means that the temperature of the walls is below the dew point or the RH is too high. In a cold climate, improving the ceiling and wall insulation should help; in a warm climate, the steps described above should be taken after showering to reduce the moisture in the bathroom.

Showers and Tubs

In a few homes I have found that very narrow grout cracks or slightly loose tiles in the wall around a tub or stall shower have led to significant leaks and decay, depending on where the cracks and loose tiles are located. For example, the amount of water that leaks into a small grout crack can be quite large if water from the tub shower flows down the crack. On the other hand, a grout crack of the same size in the wall tile at the opposite end of the tub might never be a problem. (A crack that is likely to get wet during showering can be tested by pouring water over it and looking for signs of dripping or of leakage below.)

One homeowner called me because whenever someone showered in his guest bathroom, water dripped from a drywall seam in the ceiling below. He suspected that he had a hidden leak and worried too that mold might be growing inside the ceiling. He was ready to rip out the ceiling to repair the damage.

I recommended that before hiring a plumber or contractor he make sure the water wasn't leaking from the tub as a result of careless showering. Very often, when people shower they leave the curtain hanging outside the tub, or inside the tub but partially open, and a lot of water runs down the side and through small floor and wall cracks into the ceiling of the room below. Installing a shower shield (a triangular piece of plastic) or a shower curtain liner can minimize this problem. A loose tub overflow (pop-up escutcheon plate) is another potential source of leakage and should be repaired. Make sure there is no water flowing back along the tub spout or dripping from any of the valves, because sometimes this water can flow back through an opening into the wall.

Smelly microbes can also grow in the track of a sliding shower door, and on the inside of a shower curtain, particularly at a fold, if the curtain is permanently creased. Both can be washed with a dilute bleach solution (see chapter 10 for precautions about using a bleach solution), though I recommend that a moldy shower curtain simply be replaced. (Some shower curtains can be machine-washed; see the manufacturer's directions.)

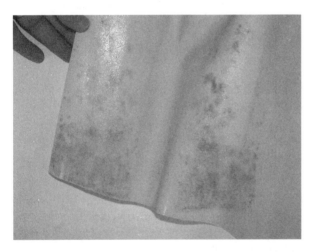

Microorganisms in the fold of a shower curtain. This plastic shower curtain was creased for months. Peeling back the crease revealed colonies of black mold. The discoloration was sampled with a swab and the swab rubbed on the nutrient in a petri dish; numerous colonies of yeast, bacteria, and black mold appeared. This mold was not *Stachybotrys chartarum*, which is mostly found growing on wet paper or wood (cellulose). The microorganisms were probably digesting soap residues and house dust.

Rooms with hot tubs often have a very strong musty odor, usually due to microbial growth on the underside of the vinyl cover (the side facing the warm water). In addition to mold and bacteria, this growth often includes a healthy colony of dust mites. Given the high humidity to which the cover is exposed, I don't know how this problem can be avoided, but if you have an odor like this around a hot tub, perhaps weekly disinfecting of the cover will help. (Contact the manufacturer for cleaning instructions.)

Sinks

Mold can grow under a bathroom or kitchen sink if leaks occur. Sometimes leaks occur intermittently inside or behind the cabinet and are thus hard to see. Be alert for any musty smells. Look periodically in and behind the cabinet, and at the ceiling of any basement or crawl space directly below the kitchen or bathroom. Fill and drain the sink; use a mirror and flashlight, if necessary, to check for leakage underneath. Water from a leaking valve can find its way into a base cabinet if the plumbing fixture is not fastened securely to the sink or countertop. Kitchen spray hoses frequently leak from the handle or from the hose under the sink, so after using the spray, check for dripping.

Plumbing Leaks

If you have had a leaking pipe behind the drywall or tile in a bathroom or kitchen, it's important to determine the extent of the mold growth before repairing the wall (and *hire professionals;* see part 3). It's likely that with any long-term leak, mold grew on the backside of the wall, so it would be prudent to remove any moldy or damp material, since in any case you don't want to paint over moldy wall materials or install tile over deteriorated drywall.

And any concealed plumbing leak in a kitchen or bathroom can lead to staining on the ceiling below, depending on where the pipe is located.

Refrigerators

Certain areas in kitchens are prone to mold growth, such as the drip tray of some frost-free refrigerators. If your refrigerator has an accessible drip tray, check the tray every two or three months and clean as necessary; keep two or three tablespoons of salt in the tray (if it is plastic), which will reduce the water activity (chapter 2) and thus deter microbial growth. A slow leak from a water line for the refrigerator icemaker is a common problem that can lead to staining patterns, warping, and mold growth in the floor under and around the appliance. Periodically check in front of, below, and behind the refrigerator for leaks (and clean out the accumulated food particles and dust, while you are at it). Mildew on the outside of a refrigerator (often the common microfungi *Cladosporium* or *Aureobasidium*) and on the door seals can be cleaned with diluted bleach. Replace a refrigerator door seal if it is cracked or broken.

Stoves and Ovens

Every time you boil water, the relative humidity in the kitchen rises as vapor enters the air. In fact, whenever anything is fried, baked, or broiled, water evaporates from the food, because just about any food contains significant amounts of moisture. In addition, when you cook with gas, water vapor constitutes about half of the combustion products from the flame. An exhaust fan vented to the exterior will carry most of the water vapor and other combustion products away. (Fans that vent back into the kitchen may remove a small amount of the grease, but they do not remove any water vapor or other combustion gases, such as carbon monoxide.)

A few words of caution: Often, the flap on the outside of a kitchen exhaust system is stuck shut by grease or paint, or the opening of the flap is blocked by a wasp nest. In addition, it's not uncommon for the installer to forget to remove a knock-out plate on the exhaust hood. So have someone stand outside when the exhaust fan is turned on, to see if the flap opens. If the flap isn't stuck but doesn't open, then the airflow is blocked. In this case you need to hire a technician to investigate. (But check first to see if there is a bird's nest in the vent pipe.)

Laundry Areas

An improperly vented dryer or a torn or loose dryer hose (a flexible duct) can introduce water vapor into a habitable space. In one home a torn dryer hose was suspended at the basement ceiling. The moisture that flowed out of the hose fueled mold growth in the lint that collected in the ceiling fiberglass, and airflows distributed spores throughout the house.

In an apartment that I investigated, previous tenants had not owned a dryer, but an existing exhaust hose had remained attached to the vent kit at the wall. The tenants had stuffed a foam sponge pad into the end of the exhaust hose, possibly to stop cold air from infiltrating. When new tenants moved in, they attached their dryer to the exhaust hose, not realizing that the hose was blocked. Unfortunately, the foam sponge pad prevented much of the hot, moist air from going to the outside, so the clothes took a long time to dry. In addition, some of the air leaked into the apartment. The new tenants used the dryer daily and wondered why it was so inefficient, and why the apartment was so damp in the winter. *Cladosporium* microfungi were growing on the wood trim and drywall near the floor on the cold exterior walls, because these surfaces were near the dew point. Microfungi were also growing in the carpeting, which was laid on concrete, and one tenant developed asthma after living in the apartment only a year.

A dryer should always be vented to the outside, never into a garage, an attic, or a habitable space, unless you want to live in a greenhouse.

And washing machines can also leak. Check the hoses frequently, using a mirror and flashlight if necessary.

Carpeting

In the Northeast I often find moldy floor dust and mite infestations in the carpeting in rooms above crawl spaces, or above cold basements or garages, particularly if the rooms face north or receive little sun. The floors above unheated spaces are usually a few degrees colder than floors above heated spaces, and thus have a higher relative humidity. In addition, if there is no vapor barrier present, water vapor from a damp below-grade space may diffuse through the subfloor into the carpet or floor dust above. In conditions of high RH, mold will grow in dust, wherever it is located (and that includes carpeting, whether hypoallergenic or not).

To minimize the likelihood of mold growth in the carpeting in such rooms, you have to reduce the RH of the air in the carpet. The simplest way to do this is to keep the room warmer and to mix the air. Both can be accomplished with a portable heater (refer to the manufacturer's directions for safe use). You can also insulate the floor (but no exposed fiberglass insulation in the basement or crawl space ceiling, please!). Moldy carpet should be replaced (see part 3).

I also discourage people from installing wall-to-wall carpeting in a kitchen or bathroom, in an entranceway, or on a screened-in porch that is open to the weather, because if any type of carpet (even all-weather or hypoallergenic) has remained wet for more than a day or two for any reason (washing, foot traffic on a rainy day), the dust within will provide nutrients for mold growth. If you insist on having carpeting in these areas, consider using area rugs or mats, which can be removed and cleaned or periodically replaced.

Potted plants on carpeting can also be a problem. If a pot is not sitting in a water-impervious tray or if someone is careless when watering, mold can grow under or around the perimeter of the pot. So unless properly protected, don't place potted plants on or near carpeting or rugs.

I am stupefied by the often-heard claim that "a carpet acts as a filter," reducing the load of particulates in the air. If you don't disturb the air in a room, many suspended (airborne) particulates will settle onto the floor, as well as onto other surfaces. If there is a wood floor, you can see the layer of dust, and if you walk by, you will resuspend some of the particles. If there is carpeting, dust will settle into the pile, and though you may not see the particles, you will still resuspend them when you walk on the fibers. And carpets, particularly those with a deep pile, can also hold a great deal of dust; in fact, such carpets can have more than fifty times the surface area of a floor without fibers, so there is that much more area (and volume) to hold particles.

When you run an air filter in a room, the air is pulled across the filter, and suspended particulates are removed by trapping. The air is thus cleaner and the filter is dirtier. When a filter is dirty, you replace it; you certainly don't walk on its surface!

Damp-mopping a vinyl or tile floor will remove almost all of the dust, but HEPA-vacuuming a carpet will remove only some of the dust and allergens present. That said, millions of homes, offices, and businesses have carpeting. With proper care and maintenance, most people have no problem with this type of floor covering. But a carpet contaminated with microbial growth, even when dry, can still release potential allergens into the air as people walk across its surface.

One woman e-mailed me this question (hard to believe, but true!): "One day I noticed a tiny mushroom growing out of the corner of the wall-to-wall carpeting in our living room. My dog ate the mushroom, and I haven't noticed any others. Is the problem solved?" Of course, the answer depended on whether she considered the dog or the mushroom the problem. In fact, the problems may have been related if the dog wasn't housebroken.

The mushroom was part of a larger fungal structure, with a mycelium that may have been spreading throughout the jute pad or even the wood beneath the carpet. I recommended that the woman hire a professional to investigate for hidden decay, the extent of which would depend on how much moisture had been supplied and over what period of time. If water had been leaking under the rug for a long time, then there was probably an extensive mold problem, and she would have to hire professionals to find and eliminate the moisture source, as well as get rid of the moldy carpet and pad and any rotted wood (under containment conditions; see part 3). A carpenter would probably be needed to repair the floor beneath. As for the dog . . .

Mold Odors

Musty odors in any room in the house can be a sign of mold growth. One family hired a contractor to add a room to the back of their house. As soon as the addition was completed, that part of the house developed a musty smell. The family wondered if a mistake made during construction had led to mold growth.

If the odor became noticeable immediately after construction, then they were probably right to be concerned, because the builder may have used moldy wood. Perhaps looking at pictures of the construction in progress would help them determine if this was the case, or they could open up a floor or wall to look at the framing. (In my opinion, lumber damaged by *macrofungi* should not be used in framing; lumber with surface *microfungal growth* resulting from damp storage should not be used in framing unless cleaned, treated with borate, and sealed.) If the odor began in the fall, however, several months after construction, then it's possible that moisture or rodents were getting into a floor or wall cavity. (And if the room had been built over an enclosed crawl space, it's possible that excess RH may have led to mildew growth and that infiltration carried the odor into the rooms above.)

One couple contacted me because they began to smell an awful odor, first in the hallway outside the kitchen, and then in the basement and master bedroom. They had looked in the kitchen cabinets and behind appliances, and had a plumber check for leaks. They hadn't found anything, and the smell just kept getting worse, especially near electrical outlets. They decided they couldn't live with it anymore, and they were ready to move out of what had been their dream house for twenty-five years.

Odor problems like these are often very difficult to solve. You have to take a step-by-step approach to determine the source or sources of the smell. The hardest sources to find are those that originate from a large surface area, such as a wood floor, carpet, wall, or ceiling. If the odor is strongest in one part of the room, then your search will be more focused at the start.

I told the couple to check if the odor was coming from electrical outlets by reducing the air pressure in the room. This is easy to do: set a box fan on exhaust in an open window and shut the doors and other windows to the room. As the fan removes room air, some air will flow back into the room from the wall cavities. I advised the couple to set up the fan and then sniff at the electrical outlets and other wall gaps. (If you are sensitized to mold, you may want to ask someone else to be the "sniffer." Also, do not reduce the air pressure in a room that contains any combustion equipment, such as a boiler, furnace, or water heater, unless the equipment is turned off, because the pressure change will cause a backdraft.) The couple subsequently discovered that rodents had moved into a wall cavity, and the odor of their nest was entering the room through a nearby electrical outlet.

It's not uncommon for rodent nests to cause such odors, particularly in wall cavities that contain insulation and are open to either a basement or a crawl space. Some rodents urinate and defecate in their nests, leaving a damp pile of biodegradable debris that begins to fester. If animals die in a wall cavity, a smell of decay can linger for weeks.

If the source of the smell is in the wall cavities, you should hire professionals to clean and disinfect the cavities under containment conditions (see part 3). In addition, it's important to find out how the animals are getting in. Pests may be using openings that are quite a distance from their nest, so check carefully at the entire perimeter of your house and garage, as well as the interior of your attic, for any openings much bigger than three-eighths of an inch. You can use wood, metal, or even masonry to close off these "doors" (be sure the pests aren't home when you shut their door!), but do not use fiberglass insulation, which is an invitation for pests to move in. Also, be sure there are no unscreened, open windows giving rodents access to the basement or crawl space.

In a room of one home, a musty odor developed after a through-wall, portable air conditioner was installed. By reducing the room pressure, I found that the odor was coming out of a narrow crack in a horizontal section of a window frame. Apparently the AC unit was not properly sealed at its perimeter, and water had started to enter the wall cavity, leading to mold growth. Another home seemed to have a mold odor throughout, and depressurization localized the odor source to a living room wall, where air flowing in from an electrical outlet had an overwhelming musty smell. Above the outlet was a single window made up of two separate units. The exterior trim between the halves of the window was leaky, and as a result, rain was able to enter the wall. Removal of a few clapboards at the exterior exposed a veritable jungle of mold, rotten wood, carpenter ants, and rodent nests!

One woman rented a new apartment and within a few days after moving in smelled mold in the bedroom closet. The manager told her there had been a roof leak, but it had been repaired and everything had been sanitized. Still, the smell got worse and worse.

I told her that mold could still be growing in the wall cavity, or it might be dead but the odor could still persist. There might also be contaminated insulation in the wall or ceiling cavity, or moldy drywall or carpet in the room. I suggested that she buy a temporary clothing rack and move all of her clothing out of the closet. If the clothes showed visible mold growth or smelled musty, she should have them cleaned and insist that the closet problem be corrected. She also needed to consider the possibility that the source of the original moisture had not been correctly repaired. Roof repairs sometimes require more than one attempt.

As discussed in chapter 4, things other than mold can produce musty odors: lamp fixtures, vinyl window screens, carpeting, paint, and even computers that off-gas. If you are desperate to figure out the source of an unpleasant smell, remove all the furnishings from the room and cover the floor with aluminum foil or Dennyfoil (paper laminated on both sides with foil; see the Resource Guide) to eliminate as many odor sources as possible. If the smell remains and the heating and AC registers are sealed, then you know that the source is probably the ceiling or wall surfaces or cavities.

A major fire in a building can leave a charred-wood smell, as well as a musty odor if water entered the wall cavities and the walls and floors were left intact. I'd be cautious about moving into a house that had suffered a major fire. Check with the local fire department about the extent of the damage and water saturation. If the house was really gutted after the fire, then all the walls and floors would have been removed and replaced, and only the framing left. Under these conditions there might not be any mold, though framing that has been charred and not properly sealed will occasionally smell, and inhabitants may complain of headaches from the odors.

Old but Good Quality?

In the 1960s and 1970s, when "back to the land" movements captured the imaginations of many young people, some moved out of urban areas and tried to live simply in the country, growing their own food and building their own houses. Some built sod huts, passive solar houses, or houses made of bales of hay. Others created residences out of old barns, or recycled barn boards by using them as paneling in their new homes.

While I think pastoral scenes are appealing in their own way, I know of one hay-bale house that had to be demolished after five years because of mold growth. And I do not recommend that people with allergies, asthma, or mold sensitivities live in old barns or in homes built partly with barn boards. In several such structures that I have investigated, the exposed boards were covered with mold and other allergens.

Soot Is Not Mold

A woman who lived on Cape Cod called me when she and her husband were about to flee their home because they had been told by an air quality "investigator" to evacuate; understandably, she was very upset. The investigator had removed a sample of black dust from a ceiling paddle fan in the kitchen and sent it to a "lab" for analysis. Based on the report from the lab, the investigator told the couple that theirs was the worst case of *Aspergillus* contamination he had ever seen.

After the wife and I had spoken for a few minutes, she told me that they burned a kerosene lantern in the kitchen every night. The lantern was placed in the middle of the table, under the ceiling fan. I began to suspect that the dust was black because it contained soot, not *Aspergillus niger* (a black mold). I suggested that they mail me a sample of the dust and wait for my call before moving into a hotel. As it turned out, the so-called *Aspergillus* mold consisted entirely of pollen particles coated with soot.

Soot can come from kerosene lanterns, gas fireplaces, engine exhaust from cars and trucks, spillage from poorly adjusted gas or oil burners, and, believe it or not, jar candles. Scented candles are very popular these days. Shelves in drug stores, supermarkets, and gift shops are stacked with them. Most jar candles unfortunately create a lot more than ambiance; they can also be a major source of soot staining in homes. If you use these candles and the jar rims are black, the candles are emitting soot.

Because of its combustion chemistry, a flame will produce more soot when disturbed. In the quiet of a room, the heated air around the flame of a tapered candle moves in one direction—upward— and the flame itself does not waver much unless disturbed. In a jar candle, the air moves in two directions: downward into the jar and upward with the hot combustion gases from the flame. This causes air turbulence, and the flame flickers, producing more soot. (Some jar candle manufacturers recommend keeping the wick short, no more than a quarter-inch long, thus reducing the flame size and soot production.)

Because soot particles are so small, you can't see them in the air, but they accumulate on all indoor surfaces. The denser the deposit, the blacker it appears. Pictures hanging on the wall protect the surface behind from soot deposits. If you take a picture down, the uncovered area may appear lighter than the rest of the wall.

Soot deposits along walls and ceilings delineate airflow differences. Air rises as it is heated by the warm surfaces it comes into contact with, such as light bulbs and baseboard convectors. When this air contains soot particles, some of the particles collide with surfaces and stick. Soot deposits are heaviest where there is the greatest amount of air striking a surface, such as on the ceiling directly above a light bulb or the wall above a baseboard convector. Black patterns from soot often appear on walls above other warm objects, such as dimmer switches or outlets with power supplies.

Temperature differences between the outside and inside of a house in a cold climate can also affect the pattern of soot deposits. All exterior walls in most newer homes are insulated to slow the loss of heat in the winter. Wherever there is wood framing (such as wall studs), the drywall is usually in direct contact with the framing, held in place by screws (or nails) driven into the wood. Because there is

Soot stains above a chandelier. The air in this room contained soot. Air was heated by the three bulbs and rose by convection in three streams above the chandelier. The soot in each air stream collided with and stuck to the ceiling, resulting in three distinct soot stains.

no insulation at those locations, and because metal conducts heat better than either wood or plaster does, the screw heads, even when covered with plaster and paint, are the coldest spots on the drywall. The next coldest wall surface will be drywall in direct contact with the wood studs. The surface of the drywall between the studs (this area is called the *bay*) will be less affected by the cold outside temperature (and will thus be closest to the indoor temperature), because the insulation is thickest toward the center of the bays. Room air very close to the wall surface of the bays will be warmer than the air equally close to the drywall fastened to the studs. But air in the middle of the room—not near walls—will be warmer still. For example, in a heated room on a winter day, the drywall at the midpoint of the surface between the studs may be 68°F, but the drywall in contact with the studs may be 65°F, and the temperature of the surface over the buried screw heads 64°F. Air in the middle of such a room may be 70°F.

The cooler air near or almost in contact with the walls will sink, and the warmer air in the middle of the room will rise to the ceiling and then flow down the exterior walls. This pattern of fluid movement is called a *convection cycle*. I believe that in the convection cycle, soot particles will accumulate where the greatest temperature differences cause the greatest air turbulence (and thus, soot problems tend to be greater in colder climates). For reasons still being debated, but perhaps because of turbulent airflow, soot is deposited most rapidly where the screw heads are located and where the drywall is in contact with the framing. Thus, walls can acquire black dots and dark stripes as soot accumulates, mirroring the airflows that deposited the particles; ceilings can acquire similar patterns.

In some homes I have investigated, jar candles were being used in nearly every room. But in fact you can burn a candle in just one room, and surfaces in another room will turn black. Soot is transported in airflows from room to room, just as odors from the kitchen can fill a house in minutes.

People have had to repaint walls and replace carpeting because of soot deposits. Some homeowners who didn't realize the source of the darkening on their walls spent thousands of dollars redecorating, only to continue to burn the candles and find they had the same problem all over again. And insurance companies have been "burned" by the cost of cleanups resulting from the use of jar candles.

Soot doesn't just adhere to walls and ceilings. If the white laminate insides of your kitchen cabinets or the white plastic interior surfaces of your refrigerator are turning black, soot from candles is the probable culprit (air containing soot gets trapped inside each time you open and close the cabinet or refrigerator door). We have received calls from desperate new mothers who were hysterical because the white plastic on the baby's formula bottle or pacifier was turning black.

Many soot problems can be solved by discontinuing the use of jar candles, if that's the source, and simply painting over the soot-stained surfaces of the home, or cleaning them if they can't be painted. And if you don't want to stop burning candles in your home after the cleanup yet you want to reduce the risk of soot deposits, choose tapered candles, but don't burn them where the flame might be disturbed regularly by air currents.

If the soot source is not candles, the cure may be much more difficult. First, you must find the source of the soot (always start by eliminating gas logs and kerosene lanterns). I have heard, but am not convinced, that soot staining on walls may be caused by motor components or fiberglass insulation in heating or cooling equipment. A poorly tuned or dirty oil-fired furnace or boiler, however, can certainly emit soot, particularly upon ignition. If you suspect that your system is generating soot, it needs to be investigated and perhaps adjusted and cleaned by a technician.

Soot from car engines (especially diesel) can also enter apartment buildings from underground parking garages through the elevator shafts. If soot is entering your apartment from the common hallway, you will often see black marks under the front door, particularly on light-colored carpeting. You might also see patterns of black staining on the doorjamb or back of the door. (Cigarettes, though they do cause yellow staining, do not produce appreciable amounts of soot and are never responsible for black discoloration on walls. Not that I would condone cigarette smoking—far from it!)

Soot can be a cosmetic problem, but is it a health problem? Some recent studies have found that when soot particles come into contact with larger allergenic particles, such as mold spores, pollen, and pet dander, they can acquire the allergens from the surface of the particulate. Should the contaminated soot become airborne again, it could be a threat to the immune system, because soot particles are much smaller than the actual mold spores, pollen, or dander, and thus can travel deeper into the respiratory system, carrying allergens with them.

How do you know that the black staining on your walls is mold, rather than soot? First, mold will lose its color when wiped with a diluted bleach solution (but so will wallpaper, so be careful), and soot will not. Second, mold growth usually has some height, whereas soot deposits are flush with the surface. Third, mold rarely grows on glass, whereas soot can accumulate on windowpanes. Take a paper towel and rub it on the inside of a window in your house; if the towel comes away black (and you have sources of soot), you'll know that the streaks on your walls may be caused by soot deposits rather than mold growth.

QUESTIONS AND ANSWERS

Question:

Our guest bathroom has a sour smell, which is sort of embarrassing. I don't see any mold on the walls or ceilings, though. Could mold be growing behind the wall and causing the smell?

Answer:

There may be mold growing in the wall cavities, but bacteria and yeast, not mold, are most likely causing the smell you describe. Rugs or mats that are frequently wet can have bacteria or yeast growing in them, and this can cause odors strong enough to make a whole house smell. Throw away any contaminated bathroom rug or mat, and in the future, hang a rug to dry if it gets damp. Bathroom mats can be placed in the dryer once a week to minimize microbial growth. Wood floors may have been soaked by urine in the past, if the previous owner was incontinent, the toilet seal leaked, or a resident pet was badly housebroken. In severe cases the flooring may have to be removed and the joists cleaned and sealed.

Question:
In the last year I have noticed that the house smells "dusty," and I have had a slight burning sensation in my nose. Any suggestions?

Answer:
The chemicals in mold odors (MVOCs) usually don't cause a burning sensation unless present in high concentrations. Other chemicals, however, such as those that sometimes off-gas from new furniture and carpeting, may cause such a sensation. Combustion products in the air make the air feel "heavy," and unburned, aerosolized fuel oil may cause a metallic taste or a sensation of burning lips. If I were you, I would install a carbon monoxide detector and have a professional such as an indoor air quality investigator check the house for carbon monoxide and other airborne contaminants.

Question:
We stayed at my cousin's house for a few days, and when we came home our clothes smelled moldy. My cousin doesn't believe that mold can be carried on personal possessions moved from place to place, but I think he's wrong. Which one of us is right? Could we have carried a mold problem back into our own home?

Answer:
Mold odors can be pervasive even when mold growth is limited or there is no visible mold growth at all. Therefore, despite the strong odor you describe, there might not be any spores on your clothing. On the other hand, if a source of mold at your cousin's house (such as a contaminated rug or couch) was disturbed, your clothing could be carrying spores or even mites, as well as odor. Still, the number of spores that you would carry back on your pants or dress would not

be much greater than the number of spores you might pick up in the summer or fall from outside air.

The real issue here is *allergy* and not *contamination*. If you are sensitized to the mold, then it's a concern, and the clothing should be washed or dry-cleaned. You would have a similar problem if your clothing picked up pet dander from someone's house and you were allergic to cats or dogs.

Question:
I am considering buying a house, but every time I visit the property, as soon as I get in the front door I notice a distinct "moldy" smell. This goes away after a few minutes as I get used to it. Is it possible to fix this problem?

Answer:
If you or anyone in your family is sensitized to mold, I'd be very cautious about purchasing a house with a musty odor, since you don't know the potential extent of any mold growth present. It's possible that the steps you could take to combat the mold would be very simple, but it's also possible that there could be a major concealed mold problem, and that any remediation required would be extensive and costly. The most prudent thing to do would be to try to determine the source of the mold odor before you purchase the house. This could mean making holes in the walls to investigate.

Question:
Year by year, our children have become sicker and sicker, all testing allergic to mold. We are looking to move soon, but I am concerned about selling our house and causing someone else to become sick.

Answer:
More and more regulatory agencies are becoming concerned about mold. In some states, seller disclosure of mold problems is required. I would strongly suggest that you solve the mold problem before selling the house; otherwise, not only may you find yourself in the middle of any ugly lawsuit, but you might sell property to unsuspecting buyers who could become similarly ill. (About six years ago I

investigated a moldy house in which an entire family was ill from mold exposures. They sold the house, and last year I was called by the new owners to determine why they were having allergy symptoms. A similar situation developed in an office I was repeatedly called to investigate.)

I admire your principles, and I encourage you to take whatever action is required to make the house safer for yourselves and for any future occupants.

TIPS

Mold can grow on furniture

- Avoid storing furniture in cool, damp spaces.
- If you have allergies or asthma, choose futon couches, and encase the mattresses with allergen covers, rather than use upholstered furniture.

Walls, ceilings, and floors can be home to mildew

- If you have visible mildew on walls and ceilings, reduce the relative humidity in your home.
- Closets with exterior walls should not be isolated from the rest of the house; leave closet doors open, or else replace solid doors with louvered doors. Install heating units in such closets or insulate the exterior walls.
- Don't let water splash directly onto a deck; install a gutter.

Rooms with water are particularly prone to mold growth

- After showering or bathing, leave the bathroom door open and air out the room with a window fan or an exhaust fan vented to the outside. To speed evaporation, you can operate a small fan to mix the air after showering. (All electrical equipment operated next to a sink in a kitchen or bathroom should be plugged into a GFI, or ground-fault interrupt, outlet.)

- If mold problems in a bathroom persist and you live in a cooler climate, consider improving ceiling or wall insulation.
- Keep the shower curtain or sliding shower door clean, and immediately clean any microbial growth at the bottom of the shower curtain or in the door track.
- Repair leaks beneath kitchen or bathroom sinks.
- If you've had any plumbing leaks in walls, employ a professional to assess the potential extent of the mold growth before repairs are undertaken.
- Carefully mark with pencil or chalk the perimeter of a ceiling or wall stain caused by a leak. Once a leak is repaired, the stain should not continue to expand. Any further spreading of the stain would indicate that the leak was not properly repaired, that a new leak has developed, or that the insulation is soaked.
- Repair any loose tiles or grout cracks.
- Keep your refrigerator drip tray clean. If the tray is plastic (not metal), keep two or three tablespoons of salt in it to deter microbial growth.
- When cooking, use an exhaust fan vented to the exterior.
- Check your washing machine (particularly the hoses and connections) periodically for leaks.
- Be sure your clothes dryer is vented to the outside.

Microfungi grow on the dust in carpeting

- Avoid wall-to-wall carpeting in bathrooms, kitchens, entranceways, and porches open to the weather. If floor covering is necessary, use area rugs that can be replaced or removed for periodic cleaning.
- Don't let carpeting get wet under or around potted plants.

Musty odors can signal mold growth

- Take a step-by-step approach to solving mold-odor problems.
- Do all you can to prevent rodents from entering wall cavities.

Chapter 8

THE SPACES
WE DON'T LIVE IN

You relax in your living room chair, you sit at the kitchen table, you read in bed at night. Your home is your kingdom, and you think you are sovereign. Well, not quite. There are other realms in the house where Nature writes the rules, and in these unconditioned spaces, spores lie waiting to usurp your control of what may be the largest investment of your life.

Attics

Many people call me because they are worried about visible mold growth in their attics. The black growth on attic sheathing and rafters consists primarily of species of *Cladosporium*, *Alternaria*, or other genera of microfungi, but rarely, if ever, the potentially toxic *Stachybotrys* mold, because attics are not consistently wet (that is, the wood does not usually have water activity above 90 percent) and thus are not usually damp enough to sustain this kind of mold.

It is common to find black mold on the attic sheathing of a gable roof (one shaped like an inverted V), especially toward the lower edges above or near the overhang, or even along one of the gable end walls. Very often the north-facing roof slope is most severely affected, because it receives little sun during the day and doesn't warm up enough to accelerate evaporation and drying out. In extreme cases you will see water dripping from the roof-shingle nails that penetrate the sheathing, which in turn will be completely blackened and even delaminated from moisture and mold. If you look down directly onto the floor beneath the nail tips, you may see stains where the drops fell.

In a cold climate, inadequate ventilation of the attic space may lead to excess humidity as moist air from the warmer rooms below infiltrates into the cooler attic around plumbing pipes, the attic hatch, or even recessed lighting fixtures. The moisture in the air then condenses on surfaces that are below the dew point. Sometimes homeowners think their houses have attic ventilation, because roof vents have been installed, yet I have seen time and time again that while the external component of the vent was properly installed, the required opening in the wood roof sheathing or soffit trim was never cut. This can be true at the ridge vent (at the peak of the roof) and at

Black mildew on attic sheathing. The nails (protruding through the sheathing between the two rafters) are what hold the roof shingles onto the sheathing. This north-facing slope of a gable roof is covered with black microfungi (not *Stachybotrys* mold). The black growth gets denser toward the overhang (at the bottom of the picture) where it is colder. The white ovals at the nail penetrations indicate a lack of mold growth; the zinc that leaches from the galvanized nails into the wood is toxic to mold.

soffit vents (at the overhang). If you hire roofers to install ventilation, make sure they cut the openings in the wood before they install ridge and soffit strip vents. (Rectangular soffit vents are fine, but
don't bother with small circular vents, which do not allow for adequate airflow.) Also make sure they install a roof vent that works.

Even with the best attic ventilation, you can still have a serious
mold problem if too much moisture enters the attic from the house.
For example, bathroom and dryer exhausts vented into the attic can
lead to condensation, particularly in the winter. If mold is growing in
your attic in just a few of the rafter bays above the bathroom or laundry area, you most likely need to vent the bathroom or dryer exhaust
to the outside (and check for leaks in exhaust hoses). If there is mildew growing on most of the attic sheathing and the attic is well ventilated, you must find and eliminate the sources of moisture. Attic
ventilation is important, but controlling the leakage of house moisture into the attic is *more* important.

If you live in a warm climate and air-condition your home, and
the AC system and/or ducts are located in the attic, moisture from
humid *outdoor* air, used to ventilate the attic, may also condense on
the outside of the system's components, such as the condensate trap
or lines, or the "suction line" to the AC coil, if they are below the dew
point of the attic air. (Very often these are frustrating problems, because condensation is intermittent and occurs only on very humid
days.) Uninsulated AC ducts (or those that leak cold air) are particularly problematic. In climates where air conditioning is used during
much of the year, the roof sheathing is too hot for condensation to
take place during the day. At night, however, if the sky is clear and
the temperature outside is low enough, the roof (particularly the
north-facing slope) may cool to below the dew point.

As discussed earlier in the book, an unbalanced heating or cooling system can also create attic condensation. I investigated one
home in New York State in which the interior side of the gable-end
sheathing behind the attic wall insulation and the exterior edges of
the attic roof sheathing were black from microfungal growth. It was
winter, and outside there were icicles hanging from the vinyl siding

at many levels on all four sides of the building. For some reason the homeowner had removed the duct for the hot-air system's only return, located in the hallway on the first floor. He had left the duct opening at the furnace, however, so all the air entering the system was coming from the very damp basement.

Since no air was being returned from the habitable rooms, the air pressure was higher in these locations than in either the attic or the wall cavities. This resulted in exfiltration to the attic and wall cavities from the habitable rooms, and moisture condensed in the attic, as well as on the sheathing behind the vinyl siding, where it froze in the winter. There was so much moisture in the walls that eventually they would have been decayed by macrofungi. Luckily this was the first winter this condition had existed. The remedy was to reinstall the return duct (and, of course, to clean up the mold and dehumidify the basement).

Check your attic periodically for mold growth. Treating attic mold with bleach isn't really effective, because the wood surfaces are porous and rough, and it's impossible to kill all of the growth. Where the mold is superficial (microfungi), there is no structural damage to the sheathing and rafters, and the growth is in the low part of one or two rafter bays, you should HEPA-vacuum the surfaces and then paint the affected wood with an alcohol-based primer, which will generally kill most of the spores and seal them into a paint film. (Consider hiring a professional to do this, because alcohol-based primers have flammable and toxic fumes, and great caution must be exercised during application. Follow all manufacturer's precautions.) Keep in mind that just painting over mold does not cure the underlying moisture problem, which *must* still be solved.

In many cases I find that one entire side of the attic is black with mold growth, but the sheathing is intact and the rafters are not damaged. In such cases a professional remediator can, under containment conditions (see part 3), clean the surfaces by soda-blasting or Dry Ice–blasting them (using baking soda or Dry Ice instead of sand, which is too abrasive). Then the wood can be sealed.

Attic sheathing in newer homes can be made of plywood or OSB

(oriented strand board, discussed in chapter 1). If, because of moisture and mold growth, plywood sheathing is delaminated and weakened, or OSB sheathing is swollen or damaged, the affected sheathing must be replaced. Of course, this means removing and replacing
roof shingles. Any rafters that are significantly decayed from macrofungal growth (most often due to leaks) may have to be repaired or
replaced as well. Occasionally the entire roof structure has to be
removed and rebuilt, though in some cases I think this work has
been done because people overestimated the significance of the
damage or because rebuilding was less expensive than professional
remediation.

Remember that attic mold growth is most often due to genera of
microfungi (*Cladosporium, Stemphylium,* and *Ulocladium,* all black,
or *Penicillium,* which may appear to be white). These fungi usually
only affect wood surfaces, whereas macrofungi destroy wood's structural integrity. If you have mold in your attic, consult an ASHI (American Society of Home Inspectors) member, an experienced roofer, or
a structural engineer for a second opinion before hiring a remediator
or tearing the house apart.

More often than not, mold on attic sheathing is not much of a
spore exposure issue in habitable spaces below. Mold of any type in
an unfinished attic may still, however, be a problem for those who
are sensitized, if the mold emits an odor or is disturbed (sometimes
people using unfinished attics as storage space disturb the growth,
or AC return ducts suck in moldy attic air through leaks and gaps).
And in finished attic spaces, microfungi can grow on carpeting or on
walls that have been dampened by roof leaks or flashing leaks
around chimney or plumbing vent pipes; mold in such spaces can
be just as problematic as mold in any other habitable room.

An Attic Odor

One couple moved into a newly constructed home after they'd had a
serious mold problem in the house they were renting, and they began to notice that they could occasionally smell the attic air in the
upstairs rooms. They found out that the air ducts between the sec-

ond floor and the attic had many unintended openings. In fact, in some spots where lighting fixtures, supply grilles, or bathroom vents were located, they could see through from the attic into the rooms below. Their daughter was mold sensitized, and they were worried that any mold spores in the attic might be carried in airflows to the rooms below.

Since the house was new, the attic was probably not very moldy. Still, allergens do collect with dust in attics, whether ventilated or not. For this reason it's a good idea to minimize the amount of attic air that enters the habitable areas below.

In another home that had a mouse infestation in the exposed insulation in the attic floor, the new owner could sometimes smell a peculiar odor in the unfinished attic. I encouraged him to hire professionals to eliminate (under containment) the old exposed attic insulation, some of which was no doubt contaminated with mold as well as rodent litter, and then to thoroughly clean out the attic, HEPA-vacuum the floor structure (in this case, the floor joists and the back of the ceiling below), have the entire floor structure spray-painted (to seal it), and reinsulate, being careful not to leave any soffit openings to the exterior through which rodents or bats could enter.

The Roof

I was asked by a management company to investigate whether there might be a connection between the owner's respiratory problems and leaks associated with the fireplace chimney (a wood-framed enclosure for a metal flue pipe). The leakage had been going on for over a year, and there were stains near the chimney on the ceiling of the second-floor bedroom (including the closet) and around the fireplace in the living room on the level below.

Apparently the wood-framed "faux" chimney had been leaking water from the cap flashing (which covered the "chimney" at the top). The metal should have been convex, but instead it was concave, so rainwater collected around the pipe at the center, rather than running off to the edges of the flashing. Over the years the puddle of water rusted the metal, and moisture leaked down around the flue pipe. There was mold on the closet drywall and in the closet carpet where mold-eating mites foraged. The flashing as well as the moldy carpeting and drywall had to be replaced.

One family moved into a new home and found a serious leak above one of the finished rooms on the top floor, directly under the roof. Part of a wall, some carpeting, and a section of the floor had been damaged. They fixed the roof and dried the carpet, but they were still worried about the possibility of a hidden mold problem, so they e-mailed me for advice.

I thought there was a good chance of microbial growth in the carpet, and possibly the pad and even the subfloor beneath, if these had been wet several times and had remained wet for a few days. Wearing a NIOSH N95 mask and operating a room fan on exhaust, someone in the family or someone they hired (if people in the family had mold sensitivities) could carefully peel back the carpet and pad to see if they had been saturated, or if there was any staining to indicate the extent of the leak and any subsequent mold or bacterial growth. I warned them to be careful not to stir up any dust. If there was significant staining or visible mold growth, I recommended that they hire an investigator to evaluate the situation, and probably have professionals eliminate and replace any moldy carpeting and pad, as well as any plywood subfloor that had significant decay or delamination. (Always use containment when very moldy materials are disturbed in a habitable space; see part 3.) If the plywood seemed intact and was just stained, and there was no odor, then it could be sealed with alcohol-based primer before being recarpeted. Finally, if the family was considering keeping the carpet, I suggested that they HEPA-vacuum up a sample of carpet dust and send it to a lab for microbial analysis before making any final decisions.

Some home inspectors walk on the roof to observe the conditions of the shingles and the chimney. I recall one inspector telling me that the last time he did this (the very last), he was moving along cautiously when suddenly the shingles and sheathing gave way with a loud crunch. He ended up with the lower half of his body dangling into the attic between the rafters. When those roof shingles were first installed, roofers were perfectly safe walking across the surface of the plywood sheathing and installing shingles with nails driven into the wood. What had weakened the roof since the installation? Most likely there had been a long-term roof leak, allowing severe macro-fungal decay.

I saw such decay during one of my home inspections, but fortunately I was in the attic when I encountered it, rather than on the roof, which happened to be flat. The attic was large because the house was huge, over eleven thousand square feet. I became suspicious when I noticed plastic sheets spread out on the attic floor and a mop with a bucket in one corner. At one location the tarpaper vapor barrier at the bottom of the fiberglass insulation batts between the roof joists was stained. I pulled down a section of the insulation, and water poured out; the roof sheathing was so rotted that the mycelia of macrofungi were in the wood. In this case the moisture originated from a leak around a poorly installed rooftop air-conditioning unit; the tarpaper prevented the moisture from evaporating. It's lucky that no AC technician ever took a step on that section of the roof while servicing the unit, because the person might have ended up in the attic.

One man e-mailed me because his roof had leaked in several spots for years before he repaired the problem. He had hoped any mold present would eventually die, since the moisture source had been eliminated. He lived near the ocean, and when the wind blew from the water, he sometimes found it hard to breathe in the rooms on the upper level of his house. Could mold growth caused by the roof leak be the culprit?

Given that the roof had leaked for a year or more before being repaired, I was pretty sure mold had grown in the unfinished attic, and even if the mold was dead, it could still remain allergenic. If the attic was ventilated, the force of the wind could stir up any moldy attic dust, which could then find its way through pipe openings and other gaps into the habitable spaces of the man's house. It's also possible that there were other sources of mold. Wind-blown rain can cause

Rotted sheathing in a flat roof. The dark plywood sheathing between the lighter-colored rafters has been damaged by wood-decaying fungi. Water poured out of the fiberglass as the insulation was lowered, and the white semicircular threadlike pattern above the knot in the rafter is probably the mycelia of macrofungi that were starting to destroy the wood structure because of the saturated conditions.

mold growth underneath the siding, for example, and even in the wall cavities of homes located near the ocean. (If the air pressure is greater in the wall cavities than in the house, mold spores can be carried indoors by infiltrating air.) I suggested that the man have an ASHI home inspector determine the extent of the problem and make recommendations for repairs.

If a roof has to be replaced, minimize the chances for exposing the sheathing to the weather. (In older houses, where planks rather than plywood or OSB were used for sheathing, rain will enter the attic through the gaps between the exposed planks.) Be aware of weather reports and know what steps need to be taken to protect the site during reroofing. If the weather is threatening, be sure the contractor comes prepared with tarps to cover the exposed roof areas. (Many roofers don't strip the entire roof at one time, so there is only a small area to protect.) If insulation is soaked by a sudden storm, it should be removed and new insulation should be installed after any damp construction materials have dried out.

When roof shingles are replaced, the pounding on the sheathing can release wood debris into the attic. Stored goods and insulation will then become covered with dust. If there is mold on the sheathing in the attic and you are concerned about mold exposure, cover stored goods with plastic sheets before the roofers start (but don't lay the plastic over recessed light fixtures in the attic floor, because this could start a fire). Isolate the attic as much as possible from the rest of the house. If there is no attic floor, be careful where you step, lest you end up in the room below. Whoever cleans up afterward should wear a NIOSH N95 mask and should HEPA-vacuum the dust on the floor or on other solid surfaces. (If new roofing has already been installed on moldy sheathing, then exposed, soiled fibrous attic insulation may have to be replaced.)

Garages

If you are not overcome by the odor of gasoline in a garage, you may be sickened by the stench of garbage from the trash barrels or the musty smell of the mold growing in a damp corner. People tend to be careless about conditions in their garages, because they think of them as outdoor spaces. Roof leaks are ignored, and moldy leaves are allowed to blow in and accumulate.

I recently visited some friends I hadn't seen in many years. As I pulled into the driveway of their immaculately maintained ten-year-old home, I noticed a streak of green moss and a long, damp-looking vertical stain on the brick veneer at the interior corner, where the two-story house met the single-story garage that projected out from the front of the house. It appeared that the garage roof had a leak.

I unloaded my suitcase from the car trunk and walked through the garage, where I could see patches of black mold on the inside drywall, mirroring the outside water stain. After I took my suitcase to the second-floor guest room, I looked out the window and discovered that my room was just above the garage roof area where I suspected a leak might be located. Before descending to the living room, I filled a glass with water, opened the window, and poured the liquid onto a suspicious-looking gap in the garage roof shingles. Water went into the opening and moments later started dripping out of the soffit trim and pouring down the brick wall where the stain was located. My host may not have liked the news I gave him about the leak, but at least he could now have the roof repaired, and they could stop bleaching the mold in the garage, which, according to them, mysteriously "kept growing back."

In this case the moldy drywall from floor to ceiling between two or three garage stud bays (and any damp or moldy insulation present) might have to be replaced. Because the garage could be opened to the outside and easily cleaned, the demolition and repairs would not be cause for much concern about contaminating the rooms in their home, as long as they kept the door leading from the garage to the house closed. I did warn them that appropriate care should be taken to protect workers and to clean up afterward.

If a garage is located below grade, conditions of high relative humidity can develop, just as in any other below-grade space, and mildew can grow in the dust on the walls. In addition, condensation can occur on the cool walls of concrete-block garages, even if above grade, particularly if they receive little sun. Goods stored up against such walls can get moldy. If a garage has drywall and the drywall touches the floor, mold can grow if water from snow melting off a car puddles on the floor and soaks into the paper.

In a detached garage that has mold in it, your exposure to spores will be minimal if you spend little time there, unless you use the space as a shop or storage area and the mold is disturbed. In most newer homes, however, the garage is attached to the side of the house or located below bedrooms or other rooms. In these cases contaminated air from moldy attached garages can infiltrate the habitable rooms. (One significant source of carbon monoxide in homes is attached garages, which is why I recommend that people avoid idling the engine inside the space, even if only for a few minutes.)

The Exterior

I often receive inquiries about "green mold" growing on the foundation wall, close to grade, but I am rarely concerned. Remember that fungi are not plants, so they do not contain chlorophyll, which is what makes plants, including algae and moss, green. Moss is not mold, but it can be a sign of excess moisture—often, splash from a leaking or overflowing gutter or leakage from a hose bibb. Moss can be scrubbed off, but be sure to eliminate the source of the water, too.

Another homeowner concern that I don't worry much about is the discoloration that appears on walls or trim on a shaded exterior of a home, even if it is mildew (dark surface microfungi). Sometimes the mildew may be subsisting on digestible components within the paint film, but more often the mildew is living on microscopic bits of organic material stuck to the surface of the paint, such as pollen, insect droppings and body parts, or the sugars in dried-out droplets of tree sap. Discoloration such as this should be washed off prior to repainting, as per the paint manufacturer's instructions.

In northern climates, such discoloration on the exterior siding will often reveal the framing within the wall, the way candle soot reveals wall studs indoors (see chapter 7). The exterior mildew grows on the wall where it is cooler, in stripes if the wall is well insulated (lightest where the studs are located and darker toward the center of the bays). The opposite pattern may occur on poorly insulated or uninsulated walls: black at the studs and clean at the bays, where the wall sheathing and siding are warmed by inside air. Though this mildew may be considered a cosmetic defect, mildew growing in certain locations on the surface of the siding or trim isn't usually a worry as far as indoor exposures are concerned.

Macrofungi on siding are definitely something to worry about, however. This growth can signal a hidden water and fungal problem beneath the siding, as well as the potential for significant concealed wood decay. In many cases the fungal growth and decay are occurring invisibly behind the siding, undiscovered until someone decides to renovate the home (or until a mushroom actually appears!).

Morning dew on clapboards. On this exterior wall the clapboards in contact with the sheathing over the studs are warmed by heat from the inside that is conducted outward through the studs (framing wood). The clapboards in contact with the sheathing over the bay cavities are cooler, because insulation in the cavities reduces heat loss from within the house. Overnight, water condensed on the cooler portions of the exterior wall. The oval dark areas on the clapboards are wet and indicate the location of the bays. The clear vertical stripes are dry areas and indicate the location of the studs within the wall. Very often mildew grows on an outside wall in patterns that mimic the areas of moisture condensation.

If the decay is widespread, it can lead to very costly repairs in both the exterior and the interior of the home, depending on the kind of construction.

Flashings

At a three-year-old condominium complex at a coastal location in the Carolinas, unit owners were astonished to find large mushrooms growing out of the siding above window and door trim beneath a large upper-level deck. In this case a minor yet significant construction error had been made on the three-level decks in each of several buildings at the complex: a metal flashing had been left out at a small gap between horizontal boards that caught rainwater. Since the day of construction, all the rainwater that collected on top of these boards flowed into the small gap and leaked down behind the siding. The vertical posts near the decks that supported the building corners, built of two-by-six lumber and concealed by the siding, were so rotted that the "wood" could be scooped out by hand.

I have investigated a variety of building failures all over the country. Most of these involved design flaws, construction errors, and inadequate maintenance that led to severe macrofungal decay of building wall components. Often it's the smallest details, such as a lack of flashing, or flashing that is tilted toward rather than away from the building, that cause the biggest problems.

Gourmet clapboards. These two beautiful mushrooms appeared at the side of a house between the time a buyer's offer was accepted and the day the home inspection occurred. There was a roof leak above the wall, and water was running down the sheathing behind the clapboards, fueling the growth of a macrofungus. As is often the case, all of the fungal mycelium and decayed sheathing were concealed by the cedar clapboards, which are resistant to fungal decay.

At a hilltop Vermont home exposed to wind-driven rain, the aluminum flashings on top of horizontal trim boards were sloped toward rather than away from the house, so rainwater flowing down the clapboards was intercepted by the flashing and trapped at the metal crease. The water then ran to the end of the metal piece, intercepted the vertical corner board trim, and entered the wall behind the siding and trim, where again much of the concealed wood was rotted.

Rotting trim board. The end of the horizontal trim board at the side of the house is severely decayed beneath the cracked paint. The pattern in the paint is characteristic of such decay. The metal flashing at the top of the trim was sloped toward rather than away from the siding (improperly installed), and water entered the wall joint between the horizontal and vertical trim. The point of the syringe needle indicates where water was entering the wall. The corner post concealed by the trim was also decayed by macrofungi.

Improperly sloped window cap flashing. The metal drip cap on top of the window trim is sloped toward the siding rather than away from it. Water accumulating on top of the aluminum flashing should have flowed off the bent edge but instead traveled to either end of the flashing. At each end of the window trim, a diagonal (mitered) joint developed a gap into which water flowed. The sheathing and framing behind the trim and clapboards around the window were severely decayed by fungal growth, and a small puffball mushroom was growing out of a gap between a clapboard and the vertical trim.

Improperly sloped drip-cap flashings are more often the rule than the exception, particularly at horizontal window and door trim boards. At many, many buildings where I have investigated decay problems, most of the hidden rotted sheathing was located around windows and doors facing the direction of wind-blown rain.

Roof Overhang

An inadequate roof overhang can also lead to trouble. I was asked to investigate one home in Florida where severely decayed sheathing was discovered when clapboards were removed during the construction of an addition. The house had a roof design that called for almost no overhang. After driving along the Gulf Coast for over two hours to reach the house, and after passing hundreds of homes along the beach, all with ample overhangs (most over two feet), I arrived at the home in question. It was only a few years old and was taller than most of the older homes I had seen; in addition, instead of having a masonry exterior, like most of the others, it was sided with painted cedar clapboards.

The home was designed so that some of the roof water flowed off the edge of the gable roof and down the siding. As I neared the building, I could see there was a scaffold up and carpenters were at work. Most of the clapboards on one side had already been replaced. On another side the clapboards had been removed to reveal severely decayed OSB sheathing. The carpenters said that around windows and doors, the sheathing was so rotted that they had been able to remove most of it with their bare hands. Not surprisingly, the cedar clapboards were not decayed at all, owing to their resistance to fungal growth.

The owner didn't think roof water running down the side of the building was causing the damage, so I asked if I could go up the scaffold and run water from a hose onto the roof. I sprayed "pretend rain" onto the shingles, and it ran down the siding until it came to a flat section of roof, where it flowed horizontally and then cascaded down another vertical wall where the clapboards had just been replaced. At the bottom of this wall was a closet accessible from the exterior. The closet had no drywall or insulation, and the sheathing was exposed. After a few minutes, water was dribbling down the interior of the sheathing into the closet. I felt sad rather than triumphant, because I knew that unless the homeowner built an adequate overhang or installed gutters, all the new OSB sheathing, even if re-sided with cedar clapboards, would rot and have to be replaced again.

Water Flow, Overhangs, and Gutters

In Wisconsin, at another newer home without any roof overhang, the homeowner had removed the gutters because they filled with ice in the winter, causing roof water to back up and get into the walls (ice damming). In this case the home was sided with vinyl over OSB. Every time it rained, roof water ran down the siding. No siding is designed to be watertight, and certainly vinyl is no exception. In fact, this type of siding has many potentially leaky gaps, particularly around windows. At this house, as water sheeted down the siding, it was intercepted by the protruding plastic window trim, and it ran behind the vinyl siding and decayed the OSB sheathing below the vertical edges of each window sill. To prevent further damage, the homeowner extended the roof overhang.

Horizontal cracks at the exterior, if small enough to attract water through capillary action, can lead to significant water entry and subsequent decay. At a home in Maine, an architect designed the trim at the roof overhang of the home's addition to resemble the trim at an exterior porch on the front of the house. The design worked fine at the open porch, because there were no exterior walls and the water dripped through cracks in the trim to the ground. At the addition, however, the water ran off the edge of the roof and down the face of the oblique trim board of the slight overhang, until it came to a horizontal joint between two boards. The opening at the joint was no wider than the thickness of a few sheets of paper, but it was enough to draw in, by capillary action, great quantities of water behind the trim. This water was then trapped within the wall and flowed down behind the clapboards, leading again to severe concealed macrofungal decay of the sheathing.

To minimize the potential for decay at the exterior of a house with a gutter system, the gutters should be kept clean to prevent overflow. At one home in Pennsylvania that I investigated, the owner hadn't cleaned the gutters for the fifteen years he had owned the house. The front gutters were clear, but in the back, where overhanging tree branches shaded the rear of the house, the gutters were clogged with leaves. When it rained, these gutters filled up quickly and overflowed. In the course of preparing the home for an addition, contractors started to remove some of the clapboards and found major concealed macrofungal decay of the sheathing and the corner posts under the trim at the rear corners.

To test what happened to the roof water after it left the clogged gutters, I filled one gutter with water from a hose. As the water poured out over the end of the gutter, which was pitched toward the rotted corner, a steady stream of liquid arced over toward the corner trim and flowed down it, soaking all the joints between the vertical trim and the horizontal clapboards. Eventually this water, drawn in by capillary action, flowed into the joints and behind the clapboards and vertical trim. Under a significant water flow, water that enters by capillary action around loose siding or a trim nail can lead to decay.

It seems rather odd, but in most of these homes there was no musty odor or other sign of extensive concealed decay. The fungi appeared to be growing on the nutrients provided by the readily decayed OSB under the mold-resistant cedar siding. In addition, the fungi were often fleshy fungi (macrofungi), which don't necessarily produce much of an odor. Because all the exterior walls contained fiberglass insulation and had vapor retarders behind the drywall, the sheathing moisture did not cause interior staining in any of these buildings. Mold growing behind or on the outside of exterior siding is less apt to be disturbed and thus seems more of an economic problem than a health issue. Still, some fungi produce MVOCs (microbial volatile organic compounds), and serious leaks can lead to significant fungal growth in structural wood behind the sheathing. Eventually the growth can spread into the wall cavities, and spores can be released into habitable rooms. So it's a good idea to prevent water intrusion, wherever it might occur, to protect your home and your health.

The Ground Around

Sometimes fungi growing in the ground around the house can have an impact on the building, as one surprised homeowner discovered. The family had lived in the house for many years and had repainted the exterior several times on a routine maintenance schedule. After the most recent painting, however, splotches appeared on the lower few feet of the siding. The homeowner complained, and the painter was forced to repaint the lower part of the house. The splotches reappeared. Further investigation revealed that the discoloration consisted of mold spore clumps from a fungus growing in the bark mulch that had been raked carefully around the foundation to improve the landscaping. Not quite the improvement the owners had hoped for! This particular type of macrofungus (artillery fungus, or *Sphaerobolus stellatus*) forcibly ejects its sticky clumps of spores into the air, and the siding was in the way.

QUESTIONS AND ANSWERS

Question:
There's black mold in the attic of my three-year-old house. How can I tell if the builder used contaminated lumber?

Answer:
If you look at the black mold, you should be able to tell whether it was on the wood to begin with or grew after the house was built. If the sheathing (the wood between the rafters) is darkest close to the lower edge of the roof (the eaves), especially the sheathing of the north- or east-facing gable in cold climates, or if the mold appears only at the top of the rafters close to the sheathing and there are drip stains below any nails, the growth is due to excess attic moisture and/or inadequate ventilation. If there are isolated areas of mold on the rafters, and no bathroom, cooking stove, or dryer is venting into the attic near that location, the lumber may have been moldy to start with, particularly if you see oval or other shaped colonies that appear to be covered and bisected by rafters.

Question:
We just added a humidifier to our hot-air heating system, because the air in the house was so dry. Now I am noticing mold in the attic for the first time. Could the humidifier be causing mold to grow?

Answer:
Yes. If there are any supply ducts in the attic, you may want to check to see if the heating system is unbalanced (as a result of gaps in the attic ductwork, for example). An unbalanced system can lead to attic condensation and mold growth. In addition, be sure that there is adequate ventilation in the attic space, and that there are no openings through which house air can leak into the attic.

TIPS

Be alert to conditions that can make attics prone to mold growth

- Maintain your roof to protect your home against water intrusion.
- Install roof vents to improve attic ventilation. Be sure the openings are sufficient (see Resource Guide).
- Do not vent bathroom or dryer exhausts into the attic or soffit. These exhausts must vent only and directly to the exterior.
- Small areas of darkened but undamaged attic sheathing can be HEPA-vacuumed and then painted with an alcohol-based sealer (exercise caution or hire a professional) if supplemental ventilation can be used to remove fumes.
- Any wood that is severely decayed, delaminated, or weakened by mold growth should be replaced by a professional.
- Air-conditioning ducts located in an attic should be insulated and airtight.
- Minimize the amount of attic air that enters habitable areas (and vice versa).
- If your attic smells musty or if you experience allergy symptoms when you enter the attic, have professionals eliminate old exposed attic insulation, and HEPA-vacuum and spray-paint the floor structure to seal in potentially allergenic dust. If you reinsulate, be careful that no openings remain through which rodents or bats can enter.
- Repair roof leaks promptly, and deal with any carpeting or walls that have been wet (see part 3).
- If your roof is being repaired or replaced, be sure the exposed work area can be protected should it rain (especially if there is plank sheathing).
- Check your attic in winter for condensation.

Don't let conditions develop in the garage that are conducive to mold growth

- Maintain your garage roof with the same care as you maintain the roof over the rest of your home.

- If there is mildew growing in the garage, reduce the moisture sources and increase the ventilation on dry days. Use a dehumidifier as a last resort, but only if your garage is fairly airtight and the problem cannot be solved by other means.
- Drywall should not touch the concrete floor.
- Whether you park your car in the garage or use the space for storage or hobbies, an attached garage should be considered part of your living space and maintained accordingly.

Do what you can to prevent exterior water from getting behind the siding

- Macrofungi on the exterior always signal hidden water and mold problems beneath the siding.
- Make sure that all the drip-cap flashings on top of exterior windows and door trim are sloped away from and not toward the siding.
- Even small gaps at the exterior can lead to water entry, particularly if they are subjected to wind-driven rain or steady flows of roof water. Check the outside of your house during heavy rain to see if water is flowing down the siding; roof water should never flow down the siding of a house.
- Keep gutters and downspouts free of plant debris.
- If your house doesn't have gutters, it should have adequate roof overhang (at least fifteen to twenty inches).

Chapter 9

TESTING FOR MOLD

Do you need to test your home for mold? If there is no visible mold in any of the rooms and no detectable musty odor, and no one in your family is suffering allergy or respiratory problems that may be mold related, then in my opinion there is no reason to sample for mold. I also don't think testing is called for if you see a small patch of mold on a bathroom wall or on a window sill, because careful cleaning can take care of the problem (unless you suspect there has been leakage in the wall cavity; refer to part 3).

Mold testing is becoming an issue in real estate transactions. Some buyers and inspectors think that mold testing should be a regular part of prepurchase home inspections, just as radon testing and pest inspections are. I do not believe that mold testing should be required in a prepurchase home inspection, for a number of reasons. First, mold test results can be very confusing (I will discuss this subject later in this chapter), and thus may not be particularly useful. Second, mold, unlike asbestos and lead paint, is a living organism and may not be present at the time of the inspection but then could start to grow and spread within days, if conditions are right. Third, there may be mold growth in rugs or furniture that will move out with the seller. And finally, the conditions under which an air sample is taken during the home inspection can have a big impact on the results.

For example, if all the windows are open, most of the spores found in the house air will probably be from the outdoors. If someone in the house recently threw out a moldy cantaloupe, spores from the fruit could be in the air. Airborne spores could even have come from the clothing of a termite inspector who checked out a moldy crawl space and then entered the house to speak to the buyer. (When I do indoor air quality investigations, which I do only for home occupants and not during real estate transactions, I make certain that every window and door in the entire house is closed and that all the mechanical systems are shut off for at least an hour before I arrive. In addition, I ask that as few people as possible, and no pets, be in the house, and that no cleaning or other maintenance work be undertaken that day, so as to minimize dust disturbances. In this way I can take air samples first with the heat or air conditioning off, and then with the system on. In addition, I can better control the release of airborne particulates from their various sources.)

I've taken thousands of dust and air samples in order to understand where and why mold grows indoors. But I believe it is much more valuable to have an experienced member of ASHI (the American Society of Home Inspectors) provide the information you need to *prevent the conditions* that can lead to mold growth than to have a fuzzy growth on a wall sampled. Chances are it's mold. Many home inspectors own moisture meters, which can help identify surfaces that have elevated moisture content, conducive to mold growth. The important questions are: Why is the mold there? What can you do about it? And how do you keep it from coming back?

Time to Test

If people are sick in an indoor environment and suspect that mold may be a cause, then testing can be useful in determining whether mold spores are in fact present. But a thorough visual inspection at the property prior to testing is still the vital first step.

I heard about one sick-building investigation in which inhabitants were having respiratory symptoms, possibly mold related. The building was large; interview questionnaires were distributed, and dozens of air and dust samples were taken. A particular type of spore, unusual for the indoors, was found in higher concentrations in one area than in other areas, pointing to a possibly contaminated heating system that was located on the roof. Investigators discovered that pigeons were nesting in the system. Unfortunately, thousands of dollars had been spent on testing to solve what was in fact a simple maintenance problem, one that could have been discovered if a visual inspection had been done before the first air sample was ever taken. Thus, I believe (and most indoor air quality professionals would agree) that the first part of any investigation is a thorough walk-through, inside and out, including a look at all mechanical equipment. This first step is as relevant for a house as it is for a larger building.

That said, it's still difficult to find consultants who test for mold in residential buildings, because it's a new field. Many of the consultants who are industrial hygienists have experience working in manufacturing facilities or office buildings, and not all are necessarily familiar with sampling for bioaerosols (particulates from living things, such as mold, mites, and bacteria). If you decide to have your home tested for mold, find a consultant in your area who is sympathetic and has had some experience with mold sampling in residential buildings.

Types of Sampling

There are three common types of sampling in indoor air quality work: *bulk sampling, air sampling,* and *surface sampling.*

Bulk Sampling

Bulk sampling involves removing a piece of contaminated material, placing it in a sterile container, and sending it to a lab for analysis. Dry samples can be sent in airtight containers, including containers made of plastic, but damp samples should not be sent sealed in plastic, because mold and bacteria can continue to proliferate in the container. Whoever does the sampling should contact the lab for instructions on collection and transport. Samples should be gathered with the utmost caution to prevent the spread of contaminated building materials and dust.

Air Sampling

There are two fundamentally different ways to sample airborne particulates, but both are called *impact sampling*, because in this type of testing, spores collide with a medium and stick to it.

The first type, *culturable air sampling*, is undertaken to identify spores that are alive and can grow on an appropriate medium. The spores are sucked by a vacuum pump into the impactor and trapped directly on the nutrient surface in a petri dish. The covered dish is then placed in an incubator at the lab and allowed to sit for a week to ten days. Only the viable mold spores germinate within and grow, often (but not always) into identifiable colonies. (Depending on the medium in the dish, bacteria and/or yeast may also grow.)

Mold colonies in a petri dish. The colonies of mold in this petri dish are nearly all circular, and each represents a point where one or more spores impacted on the surface of the nutrient agar during an air sample taken in a basement with a dirt floor. Many of the colonies were *Aspergillus* mold. A comparable three-minute air sample at the outdoors had a tenth as many colonies and very little *Aspergillus* growth.

One drawback to culturable sampling is that even though there may be many spores of one particular type in an air sample, if they are dead they will not produce any colonies at all, yet the dead spores can still be allergenic. (In fact, an indoor environment may contain over fifty times as many dead spores as living ones.) In addition, even though they can cause allergy symptoms, most hyphae do not grow into colonies on a petri dish.

Another drawback to culturable sampling is that aerosolized spores often exist in clusters of a dozen or more. But when such a cluster lands in a petri dish, no matter how many spores it contains, the resulting growth will appear to be only a single colony. (This is why lab results are reported in terms of colony-forming units per cubic meter of air rather than spores per cubic meter of air.) Thus, sampling for only culturable spores may seriously underestimate the actual concentration of spores and the exposure to allergens.

Culturable sampling also will not tell you whether the spores in the environment contain mycotoxins, because in the petri dish, the fungus may not produce toxins. If you suspect that you have been exposed to mold toxins and you really want to know, hire a professional to collect a bulk sample and send it to a lab for mycotoxin analysis, because petri-dish culturing of mold spore samples alone will not suffice. Keep in mind, though, that even if you find mycotoxins in the dust, you still will not know the extent of your exposure or how it may be affecting your health.

There are homeowner kits that use nutrient petri dishes. In this type of culturable testing, called *settle-plate testing,* a petri dish is left open in a room for up to an hour and is then sent off to a lab for analysis. Although in very contaminated environments this type of testing may provide useful information, I don't generally recommend it because of the great uncertainties involved. For example, the number of suspended particles will vary with airflows, indoor human and pet movements, and outdoor activities near the house (construction or demolition, for example). It's therefore difficult to connect the growth that may occur in the petri dish with the sources of the spores, either indoors or outdoors. In addition, the settle-plate test is biased toward larger spores, because they settle out of the air faster than smaller spores do. Yet it is the smaller spores that may be more problematic for sensitized individuals. And unlike impact samplers, which use pumps that can be calibrated for air-flow, settle-plate testing cannot provide any information regarding the concentration of spores, so this type of sampling is useless in determining exposure levels.

In the second type of air sampling, called *spore-trap sampling,* air is drawn through a device by a pump or blower, and suspended particulates, including both viable (alive and culturable) and dead spores, as well as hyphae, are trapped not in a petri dish but on a sticky surface such as grease or tape. The samples can then be stained and examined under a microscope. In my own investigations I have found that spore-trap sampling is usually adequate to determine if there is a mold problem in a building, though any air sampling done indoors, whatever the type, should also include one or two samples of outdoor air for comparison (as "controls"), unless the season is winter and the ground is frozen or under snow cover, in which case there are few spores outdoors. Unfortunately, a consultant is needed to undertake spore-trap sampling, and not all consultants do this kind of testing.

Surface Sampling

In sampling of surfaces, settled dust or growing mold is removed, either by a vacuum device or with sticky tape, or wiped with a dry swab, and observed under a microscope. In another method of surface sampling, a damp cotton swab is rubbed on a surface and placed in a sterile container. The organisms on the swab are then transferred to a nutrient and incubated.

Most of the airborne allergenic particles indoors originate in the dust on or in contaminated sources, and the air is the primary vehicle for transporting the allergens. Although some experts are very concerned about concentrations in the air, in my opinion the most straightforward means of determining whether you have a mold exposure problem is to sample the dust. Air sampling is expensive and is generally undertaken by professionals, whereas most sampling of carpets, cushions, pillows, and some components in heating and air-conditioning systems can be done by the occupant and the samples sent to a lab. Many labs (see the Resource Guide) will accept sticky-tape samples from homeowners, and the cost of sample analysis by microscopy is usually under $100.

Testing Results

If you are considering hiring someone to do mold sampling in your home, find out how the samples will be taken, and look into the education and experience of the person doing the sampling. If the particular fungal species in the samples are to be identified, find out if the samples will be sent to a qualified environmental lab for analysis. (An organization called the American Industrial Hygiene Association, or AIHA, has a program called EMPAT—Environmental Microbiology Proficiency Analytical Testing—which tests and certifies labs that provide quantitative testing.) Finally, ask what kind of information and explanation will be included in the report. Take all of these steps *before* you make a commitment.

Identifying Species

You may need to know the fungal species, as well as their concentrations, for medical or legal reasons. In such cases, air and bulk sampling for culturable mold should be professionally undertaken. On the other hand (and here mycologists and other indoor air quality professionals will probably *disagree* with me), I believe that determining the species of fungal spores present in a home is rarely necessary for purposes of protecting your health, since *any* spores, whether "toxic" or not and whether alive or dead, can cause symptoms if present in unusually high concentration.

If the report you receive has a long list of the fungal species found in your home, what does it mean for your health? The genera of the most common molds found indoors are *Cladosporium*, *Penicillium*, and *Aspergillus*. Others may include the genus *Alternaria* or even *Stachybotrys*. Reading the list of fungal species in a report is a little like looking at the pixels (dots) that make up a TV image, or looking at a graphical analysis of the individual frequencies that make up the sound from an orchestra. From a scientific perspective, these segments are interesting, but they don't present the whole picture in a recognizable pattern.

I prefer visual methods (such as direct microscopy of bulk samples and spore-trap samples gathered in various areas of the home), rather than testing that only counts the concentration of culturable spores (or colony-forming units), because microscopy provides a more complete story of potential allergens in the home. For example, whereas the surfaces of spores from outdoors are usually clean, mold spores sampled indoors in buildings with hot-air heat or central air conditioning are sometimes coated with accumulated microscopic particles of soot or paint pigment, which are smaller than spores. (When mold spores have been present in an air conveyance system, particles such as soot and paint pigment carried on airflows can impact and stick to the spore surfaces.) In addition, in samples of air exiting a heating or cooling system I very often find mold hyphae and spores wrapped around fiberglass fibers. This tells me that mold is growing in the fiberglass lining material in the system.

Concentrations

A test result sometimes includes the total concentration of spores, living and dead, expressed as the number of spores per cubic meter of air. Although outdoor spore concentrations can vary from hundreds of thousands (in the forest) to millions (in the midst of crop harvesting), total concentrations indoors usually vary from fewer than 100 to over 10,000 spores per cubic meter of air.

According to the American Academy of Allergy, Asthma, and Immunology (AAAAI), fewer than 6,500 spores per cubic meter of *outside* air is considered a "low" concentration, and above 13,000 is a "high" concentration. Concentrations between low and high are considered "moderate," and a concentration above 50,000 is considered "very high." According to the AAAAI, at the "very high" level nearly all sensitized individuals will experience symptoms outdoors, whereas only the highly sensitized will experience symptoms at the "low" level. The AAAAI also says that at the middle level of "moderate," about 10,000 spores per cubic meter, many sensitized individuals will have symptoms.*

Since there are 1,000 liters of air in a cubic meter, the middle level of "moderate" represents a concentration of 10 spores per liter of air, or about 5 per breath (assuming about 0.5 liter per breath). Many of the contaminated homes that I have investigated had moderate concentrations of airborne spores, particularly in basements, both finished and unfinished.

In my own experience I have found that just starting to draw a breath of "contaminated" air can cause me to cough, even in a space with few particulates of any type but containing fewer than 500 spores per cubic meter of air. At this concentration there is, on average, 1 spore per 2 liters of air, and therefore there are likely to be no spores in a fraction of a breath. Other very sensitized individuals describe similar immediate responses upon entering contaminated spaces. To me this suggests the likelihood of particulates smaller than spores that may be serving as surrogate carriers of mold allergens.

For example, as noted earlier, any mold-colony micro-particles (see chapter 4), as well as any house dust, plaster, or soot particles that have been in close contact with a fungal mycelium or spores, could be made airborne when the dry colony is disturbed, thus creating an aerosol containing fungal enzymes and other allergenic metabolites. Whether using culturable or spore-trap sampling methods, these allergens would not be "seen" and could only be detected using very sensitive chemical tests that recognize specific allergens. Thus, a test showing an absence of mold spores does not necessarily mean that allergens and toxins are not present in the air.

An acceptable level for the total concentration of fungal spores (again expressed as cfu/m^3, or colony-forming units per cubic meter, rather than as spores per cubic meter) in indoor air has not been

_____

*Currently posted at www.aaaai.org/nab/index.cfm?p=reading_charts.

In manufacturing facilities where *subtilisin* (a bacterial enzyme used in most laundry detergents) was first processed, up to 50 percent of the workers were sensitized to the enzyme and many developed occupational asthma. As a result of these experiences, great care is now exercised in formulating these detergents: the bioaerosol concentration of subtilisin in some facilities is maintained below 15,000 trillionths of a gram per cubic meter of air.

Some fungal protein enzymes, such as proteases, are related structurally to subtilisin. A single mold spore could weigh about 300 trillionths of a gram and contain about 10 trillionths of a gram of enzymes; at an indoor concentration of 1,000 spores per cubic meter of air, about 10,000 trillionths of a gram of enzyme might be present—very close to the indoor "safe" workplace limit of subtilisin. It does not seem unreasonable, therefore, to assume that allergic sensitization to mold enzymes can occur at concentrations of spores found in homes contaminated with the growth of common microfungi. (In some homes I have tested, the concentration of spores in the basement or coming from the heating system was over 5,000 per cubic meter of air.)

established for culturable sampling with petri dishes, nor have acceptable concentrations for individual species been defined. Nonetheless, mycologists have provided some helpful information. For example, in almost a thousand indoor air samples taken all over the country and sent for analysis to Mycotech Biological, Inc., in Jewett, Texas, Larry Robertson noted that the concentration of culturable spores ranged from zero to over 6,000 cfu/m^3, and that the average was 157 cfu/m^3.* The concentration in 87 percent of the samples was

*L. D. Robertson, "Monitoring Viable Fungal and Bacterial Bioaerosol Concentrations to Identify Acceptable Levels for Common Indoor Environments," *Indoor Built Environ* (Mycotech Biological, Inc., Jewett, Tex.), vol. 6 (1997):295–300.

under 300 cfu/m^3, and only 7 percent had concentrations between 300 and 500 cfu/m^3.

If you have culturable testing done in your home and the spore concentration is well in excess of 300 cfu/m^3, you are in that 7 percent, and you may have a mold problem. (Of course, as always, indoor concentrations should be compared with those outdoors, which are usually higher.) In addition, Robertson suggested that indoors, an individual fungal species should not contribute more than 50 cfu/m^3 to the total spore concentration, with the exception of *Cladosporium* spores (the most common outdoor fungus and thus frequently found indoors).

In *outdoor* air, the species and spore concentrations vary according to the season and weather conditions. On a windy day the type and concentrations of spores can even change from moment to moment, as moldy leaves are blown around and spores are carried on airflows. During the spring, summer, and fall there may be thousands of spores per cubic meter of air, whereas in the winter there may be few if any.

Mold species and spore concentrations can vary *indoors* too, because spores enter buildings through open windows and doors or on people's clothing or shoes, but if molds are growing in carpeting or in the heating or cooling system, the spores or allergens from the spores will almost always be present. Although these spore concentrations may vary drastically, depending on whether the mold has been disturbed, as long as indoor conditions remain the same, the varieties of mold that grow indoors will not change that much. Because we spend so much time in buildings, our exposures to the limited types of spores from molds growing indoors can be fairly constant. I believe, therefore, that people are more likely to develop an allergy to indoor mold in a contaminated building than to develop an allergy to outdoor mold. And the more sensitive to mold the people in your household are, the more worried you should be about even low levels of spores indoors, particularly if the levels are consistently greater than the levels outdoors. And, of course, the presence of a potentially toxic species in an air sample taken in a room that is highly trafficked is reason for heightened concern.

On the other hand, if no one in your family is sensitive or allergic to mold and the reported concentrations are low, there is less reason for concern. In addition, finding mold through bulk sampling of an isolated area (for example, the corner of an unfinished basement in a house with hot-water heat) is also not as serious. Test results, including reported concentrations, should thus be considered in light of potential exposure.

Finally, it seems to me that reporting airborne concentrations at any given time does not necessarily give an accurate picture of inhabitants' potential exposures. For example, if a favorite couch contains growing mold (uncommon, but something I have seen in many "sick" houses), you will likely inhale spores when you sit on the cushions and aerosolize contaminated dust. Yet if no one has compressed the cushions, that room may have low concentrations of airborne spores at the time of testing. Exposures to very high concentrations of spores, hyphae, and fungal micro-particles can be intermittent, depending on whether a source of contamination has been disturbed.

QUESTIONS AND ANSWERS

Question:
I bought a settle-plate home mold test kit. The lab results came back saying there was *Penicillium* mold in every room (the living room had eleven colonies) and a lot of *Stachybotrys* mold in the bedroom. What do you think I should do next?

Answer:
The recent rapid increase in the number of consultants investigating mold and of laboratories providing analyses has led to many misleading reports. For example, most indoor air quality professionals think that the settle-plate test you used is generally a poor indicator of indoor mold problems. In addition, some of the labs are using rel-

atively unqualified technicians who may not necessarily have the experience to identify samples accurately. (In one case I was involved in, the lab found six molds, only one of which was actually in the tape sample.)

You told me that your test revealed *Stachybotrys* mold in the bedroom. I've taken thousands of air samples in residential spaces, and very few contained this mold. Of those that did, none contained more than a few spores. So the first step to take is to have testing done by professionals and the results analyzed by a qualified lab. If this second test confirms your original results, then you definitely have a mold problem! In that case, hire professionals to find and eradicate the sources of contamination under containment conditions (see part 3).

PART III

THE CLEANUP

Many people call our office because they are afraid they may have contaminated their homes with spores while removing a moldy carpet or while cleaning mildew from a bathroom wall or ceiling. In this final part of the book we discuss ways to clean up mold and to minimize the spread of moldy dust and debris in smaller-scale as well as larger-scale mold cleanup jobs. There is a big difference, however, between moldy dust, which can be handled in many cases

A mold remediation project. Though only their plastic suits and booties are visible, these professional mold remediators were also wearing hoods and full-face respirators for protection. They are standing on a floor structure that was built over a dirt crawl space. Wood-destroying fungi, visible on the board in the lower right-hand corner, grew in the damp crawl space and silently consumed much of the structure.

by home occupants, and mold growth (particularly the kind that occurs after flooding), which is much more serious and which can require professional remediation (removal and cleanup). Chapter 10 deals with smaller-scale mold cleanup tasks. Chapter 11 discusses mold remediation done by professionals.

There are a number of publications on mold cleanup and remediation methods, some of which are listed in the Resource Guide at the end of the book. We urge you to consult these resources before taking on *any* mold cleanup work yourself. We are not professional mold remediators, and the information offered in chapter 10, as well as in certain sections of chapters 5 through 8, is intended to be used as a guide, not as the ultimate authority.

When it comes to mold growth, the big question facing homeowners is: Should I clean it up myself, or hire someone to do it? The answer depends not only on the size of the job but also on the sensitivities of family members to mold. We cannot state it strongly enough or often enough: *mold should be handled with the utmost caution, the work area should be isolated (contained) as much as possible from the rest of your home, and if you or anyone in your family has mold sensitivities, hire professionals to handle all mold remediation work.*

Chapter 10
SMALL-SCALE CLEANUP JOBS

Some of you may have cleaned mildew from a bathroom wall or ceiling, or mold from the outside or inside of your refrigerator. No doubt most of you have found moldy vegetables or bread and tossed these items into the garbage without a second thought. Some of the smallest mold cleanup jobs have been handled time and time again by home occupants, but if you or anyone in your family has mold or other environmental sensitivities, other people may have to handle what might seem to be minor mold cleanup tasks.

If you think you are not sensitized and you want to tackle a small mold cleanup job yourself, wear protective clothing, including a NIOSH N95 mask, and set up containment-like conditions (described later in this chapter) to isolate the area from the rest of your home. If you feel any ill effects while eradicating what seems to be minor mold growth, stop working immediately, call your physician, and arrange to have professionals finish the work. Finally, keep in mind that mold is a living organism and will grow in any suitable environment. Cleaning efforts will not lead to a long-term solution unless you eliminate the conditions that led to the mold problem in the first place.

The precautions described in this chapter may seem excessive to many people, but even small mold exposures can cause allergic reactions or in rare cases anaphylaxis for those who are highly sensitized. Significant exposures to some molds can make *anyone* sick. It is therefore always best to be overly cautious when dealing with mold.

The Size of the Cleanup Job

The U.S. Environmental Protection Agency (EPA) divides mold remediation into three levels: "small," where the total visible surface area affected is less than ten square feet; "medium," where the area is between ten and one hundred square feet; and "large," where the affected area is greater than one hundred square feet or where exposures during mitigation are potentially significant.* The New York City Department of Health and Mental Hygiene, Bureau of Environmental and Occupational Disease Epidemiology, describes five levels of contamination:

- small isolated areas (level I), ten square feet or less
- midsized isolated areas (level II), from ten to thirty square feet
- large isolated areas (level III), from thirty to one hundred square feet
- extensive contamination (level IV), greater than one hundred square feet
- remediation of HVAC systems (level V)†

It is my view that even if no one in your family has mold sensitivities, in the interest of health, any areas requiring mold remediation that go beyond "small" should be evaluated and probably handled by experienced professionals (see chapter 11). Deciding which description is applicable can sometimes be tricky, however. For example, it can be difficult to distinguish between superficial mold growth on a small area (what people call *mildew*) due to excess relative hu-

Mold Remediation in Schools and Commercial Buildings (Washington, D.C.: Environmental Protection Agency, 2001).

†*Guidelines on Assessment and Remediation of Fungi in Indoor Environments* (New York: New York City Department of Health and Mental Hygiene, Bureau of Environmental and Occupational Disease Epidemiology, 2000), pp. 8–10.

midity and surface mold growth due to leaks or floods, where minor visible growth could signal significant concealed growth within wall or ceiling cavities. If you see leak patterns, there's a good chance you have a concealed mold problem. In this case, what looks like a small job might be more appropriate for professionals to handle.

If you suspect that a wall or ceiling stain is damp but you can't tell for sure by touch, you can have an ASHI (American Society of Home Inspectors) member test the area with a moisture meter or buy a meter from an equipment supplier and test the area yourself.

If you have any questions about the potential dimensions of a mold problem, it's best to err on the conservative side and assume the worst.

Floodwater Categories

The IICRC, or Institute of Inspection, Cleaning, and Restoration Certification (see the Resource Guide), recognizes three categories of water that are also useful yardsticks when considering how to handle flooding:*

- clean water (category 1), from rainfall or water supply pipes
- gray water (category 2), from a washing machine, an overflowing toilet (urine, no feces), or water that has flooded over a very dirty carpet
- black water (category 3), from a sewer backup, or river water containing silt, organic debris, or dangerous chemicals

I think any cleanup job that involves category 2 or 3 water should be handled by professionals; in fact, professionals should even deal with category 1 water when the carpeting, furniture, walls, or other surfaces have been saturated for more than twenty-four to forty-eight hours, because microbial growth is so likely. And remember, the warmer it is, the less time it takes for mold and bacteria to prolif-

---

*Standard and Reference Guide for Professional Water Damage Restoration S500-94 (Vancouver, Wash.: Institute of Inspection, Cleaning, and Restoration Certification, 1995), p. 8.

erate, so moving quickly is crucial. (Flooding is discussed in chapter 11.)

Containment-Like Conditions for Smaller Jobs

While it isn't always possible for homeowners to create the containment conditions that professional remediators create, it's important to set up containment-*like* conditions to isolate the work area as much as possible. Clear the room of uncontaminated furniture and rugs; items that can't be removed can be sealed in plastic. Protect the floor with a nonslip covering. Tack or tape (using tape that will not damage the surfaces) two pieces of plastic sheeting to the top of the wood trim over entrances to the room to cover any *open* doorways. Each piece should cover approximately two-thirds of the opening. Attach one piece to the top and right side of the doorway, and the other piece to the top and left side of the doorway. The section in the middle where the two pieces overlap will function as a partial air seal while also allowing you to enter and exit. If the entrance to the room is wide and doesn't have a door, place a spring-loaded curtain rod at the top, across the opening, and tape the two plastic sheets over the rod to isolate the space.

Alternatively, you can create a "booth" around your work area by arranging four spring-loaded rods vertically between the floor and ceiling and attaching horizontal pipes to the rods with hose clamps or duct tape (see Aramsco, Inc., in the Resource Guide). Then you can tape plastic sheets to the pipes to enclose the space. Again, leave an overlap between two of the pieces of plastic so you can enter and exit. *In designing any kind of containment-like conditions, be sure that one wall of the isolated area has a window to the exterior, so a box fan can be used on exhaust.* This will remove many of the spores that may become aerosolized when the mold is disturbed. (Also be sure that the containment-like arrangement lets in air to replace the air being removed by the exhaust fan.)

In most cases, after the mold growth has been eliminated, and with the fan still operating on exhaust, other room surfaces can be cleaned with a HEPA (high-efficiency particulate arrestance) vacuum to eliminate contaminated dust; nonporous surfaces can be damp-wiped and allowed to dry.

Keep in mind that this nonprofessional, containment-like setup should be used only for small jobs. In chapter 11 we describe the containment measures that professionals use. (If you have had a leak in your home, be careful from the start to set up containment-like conditions when someone is removing ceiling and wall materials, because the debris may be contaminated with mold. Be certain that all debris and dust have been HEPA-vacuumed and surfaces damp-wiped before the containment-like enclosure is removed.)

Musty Odors

If you have a musty odor in your home, don't mask the smell with fragrances or attempt to eliminate the odor with ionizers or ozone machines (ozone is an irritating gas, even in low concentrations). Find the source of the odor, and if it is mold, eliminate the growth as well as the conditions that fostered it.

Some people consider purchasing a portable HEPA air cleaner to solve a mold problem. If there is mold growing in carpet dust and you are not in the room, an appropriately sized HEPA air cleaner will probably remove most of the spores suspended in the air. But if you walk into the room and disturb the moldy dust, the airflow from the cleaner may blow the spores around faster than it can remove them, and you will experience an increased exposure. Furthermore, if you have mold growing in your pillow, a HEPA air cleaner in every corner of the room still won't make you feel any better, because as long as your head is on the pillow, the spores will get to your nose before they get to the filter. On the other hand, once you have eliminated the source of the spores, the portable HEPA unit will help to keep your indoor air clean.

Stored Goods

Possessions stored up against foundation walls and cardboard boxes set down directly on below-grade concrete floors or stored in cool, damp spaces can acquire fungal growth. Discard all boxes that are moldy or have been wet. Evaluate the contents of each box, and clean the items outdoors or discard them. (Items with solid surfaces can probably be cleaned, whereas items with porous surfaces may have to be discarded.)

Even small, easily moved items can create spore exposures if they're moldy, so wear plastic safety goggles and a NIOSH N95 mask.* If you have a beard that interferes with the "face seal," you must wear a full face mask or have someone who is clean-shaven do the work. Consider wearing a disposable Tyvek® suit and latex or nitrile gloves (wear nitrile gloves if you have a latex allergy). Don't wear the same clothing and shoes back into the habitable areas of your home.

Do all you can to avoid spreading moldy dust from a basement or garage to the upstairs or any adjacent finished rooms. Go in and out through the garage door or bulkhead, and seal any doors leading to habitable spaces. If you are not able to use an exit that leads directly to the outside and moldy items have to be removed through habitable areas, bag the items in plastic first, and if at all possible seal any doors that lead to other rooms, to isolate your path. Remove hall rugs or protect carpeting with a nonslip covering. If you are removing a number of moldy boxes or pieces of furniture stored in your basement, place an exhaust fan in a basement window to reduce the air pressure in the space and reverse the airflow from the basement into habitable spaces. *Do not depressurize a room that contains operating combustion equipment (such as a boiler, furnace, or water heater) or anything with a vented flame (fire in a fireplace or wood stove), because this can cause backdrafting.*

Books and Papers

If any of your books and papers have a musty smell but no visible mold, take them outside. Carefully HEPA-vacuum any dust, and air

---

*For details about respiratory protection, see OSHA 29 CFR 1910.134, available at www.osha.gov/.

the items out in the sun to see if this makes a difference. Any replaceable books and nonessential papers that have been damp or that contain visible mold, however, may have to be discarded.

If you own a book collection that you value, I do not recommend that you store it in a basement (where high humidity and condensation can readily occur) or in the attic (where the heat can accelerate the degradation of the cellulose in paper). It's also a good idea to identify *in advance* anyone to whom you might turn in the event of a flood. For example, before mold begins to grow (within hours after books and papers have gotten wet), the items can be frozen and sent to a facility where they can be defrosted and then dried quickly under desert conditions. Books with mildew can be HEPA-vacuumed (thoroughly and carefully) and "sanitized"—work best done with professional guidance. (Contact a restoration company in your area, or see the contact information for Northeast Document Conservation Center in the Resource Guide at the end of the book.) Keep in mind that even the most thorough cleaning may not remove all of the mold allergens.

Clothing

Sometimes shirts or dresses that have been stored in a musty-smelling closet don't seem to be badly affected; nonetheless, the clothing should still be washed or dry-cleaned, or at the very least, in the absence of any musty smell, tumbled in a dryer, which will remove and blow away some of the moldy dust that may be present. Any clothing that has a moldy smell or minor visible mold growth should definitely be washed or dry-cleaned and then hung in the sun. If the mold growth is extensive, or if several washings and airings do not make a difference, the clothing may have to be discarded. Finally, before carrying moldy clothes through the house, place the items in a plastic bag after removing them from a bureau or closet.

Leather shoes may acquire light mildew growth that can be removed (outside) with a soap appropriate for leather, and the shoes can then be polished. Shoes or other leather goods that have extensive mold growth may have to be discarded. Check first with the manufacturer or a cleaning professional familiar with the care of such items.

Curtains and Cushions

The insides of window curtains often get moldy at the bottom edges, because they get damp from water condensing on windows during the winter, or because they rest against the cool window surface and are exposed to conditions of high relative humidity. Thin curtains or other cloth window treatments can be laundered or dry-cleaned according to the manufacturer's instructions, but again, if the items still have an odor after cleaning and airing cycles, consider replacing them. Mattresses, pillows, bedding, stuffed animals, and cushions that have visible mold growth or a strong musty odor should be bagged in plastic and thrown away. The same can be said of moldy-smelling or discolored fleecy items such as blankets that have been stored in a damp basement or unheated garage. *Remember to wear a NIOSH N95 mask and protect the surroundings when disturbing such moldy items.*

If you can't afford to replace a mattress that has been only *slightly* damp for a few hours and has *no* musty odor, you can dry it out completely in the sun (put the mattress on a table or milk crates, and *not* directly on the ground), and then encase it in a plastic mite allergy cover. This cover will prevent allergens from reaching you. If the mattress doesn't dry out within a few hours, it should be discarded. (I do not recommend saving a mattress that has a moldy odor.) Sleeping on an uncovered festering mattress can create problems, as one couple discovered. They bought what they thought was a new futon mattress for their infant's room, then called me because the child developed severe allergies. My sampling revealed that the stuffing was full of pollen and was infested with mites and growing mold. I suspect that an old mattress had been left outdoors, placed in the rubbish, and then retrieved, covered with fresh ticking, and sold as new—an illegal and unconscionable practice!

Small Surface Jobs

Small patches of mildew on solid nonporous surfaces such as painted walls and ceilings can be wiped with diluted bleach—about one part bleach in ten parts water (never use undiluted bleach for cleaning). When you are working with a solution of bleach and water, wear protective gloves and eye protection, remove furniture as needed (or use plastic to cover surfaces that may be damaged by bleach), and place a box fan in a nearby window on exhaust, to reduce the air pressure in the room. Do not mix bleach with liquid soap containing quaternary ammonium detergents, or with ammonia, because a toxic gas may be formed when these ingredients are combined. Follow directions carefully, as with any cleaning product. (Sometimes the bleach will combine with the soil or stains you are cleaning to produce toxic gases, and this is why you should use diluted bleach on a larger surface only if there is plenty of ventilation present.)

Be aware that bleach can affect surface color, so always test a small area first, and avoid using a chlorine bleach solution if anyone in the household is sensitized to chlorine compounds. (A bleach solution is not necessary for cleaning, and some government agencies now discourage the use of bleach. Moldy surfaces can also be wiped with a solution of any cleaning agent intended for the type of surface being cleaned.) If the solution you use doesn't harm the surface, wait a few moments before rinsing, to maximize the antimicrobial effect. Then rinse and allow the surface to dry thoroughly, to avoid developing another mold problem.

When it comes to moldy drop or acoustical ceiling tiles, while one or two slightly discolored tiles can be very cautiously removed by a homeowner (wearing safety goggles, protective clothing, and a NIOSH N95 mask, and working in containment-like conditions, defined earlier in this chapter), large numbers of moldy tiles should always be professionally removed.

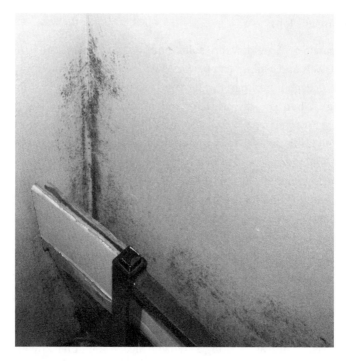

Mildew on a cold exterior wall. In winter, black *Cladosporium* mold grew behind a child's bed on the cold exterior walls of the bedroom, because the relative humidity in the apartment was too high—about 55 percent. Mites were also abundant because of the excessively high relative humidity, and the child suffered from asthma. In this case the mildew was growing in dust on a solid, painted plaster surface and covered less than ten square feet of the wall—a "level I" cleanup job. The baseboard convectors (including the fin tubing) also had to be cleaned of all dust. (Steam vapor machines, discussed in this chapter, are useful for cleaning fin tubing.)

Furniture

If you find mildew on the *unfinished* back or underside of a piece of wooden furniture, carry the item outside carefully, so as not to disturb the mold growth. To eliminate the mildew, wipe the moldy surface with a rag moistened with denatured or isopropyl (rubbing) alcohol. Wear gloves and eye protection, refer to manufacturer's instructions, and be cautious, because the fumes should not be inhaled, and the alcohol and any surface soaked with alcohol are combustible. A NIOSH N95 mask offers no protection against harmful vapors but should be worn to protect against inhalation of spores. Only work with alcohol outside. If you *have* to work inside because the piece of furniture is too heavy to remove from the house, be sure you have lots of ventilation, set up containment-like conditions, operate an exhaust fan, and keep a fire extinguisher handy. Wear at least a half-face respirator that removes organic vapors as well as spores.

Once the mildew has been treated, seal the wood with a thin coat of clear shellac or varnish. If you use oil-based varnish, wait for the rubbing alcohol to dry completely. (Rubbing alcohol is 70 percent alcohol and 30 percent water, so the wood will still be moist after the

Aspergillus mold on the front door. A couple and two children with asthma were living in a split-level home that was kept too cold in winter. Nearly colorless *Aspergillus* mold grew on the lower, and cooler, half of the door (the flash of the camera revealed the growth). Every time someone left or entered the home, spores and microparticles were aerosolized. A wood or metal door with a small amount of such growth can be cleaned as if it were a piece of furniture, either indoors under containment-like conditions or outdoors.

alcohol dries.) The advantage of shellac is that it can be applied to the wood while it's still slightly damp; in addition, shellac dries faster than varnish. Do not get any alcohol or shellac on the painted or varnished areas of the wood, as these may damage the finish. And *never* use combustible liquids such as alcohol or shellac near open flames or near anything that can produce a spark. Do not place rags soaked with alcohol, varnish, or shellac directly in the trash; allow them to dry first.

Mildew on *finished* (painted or varnished) furniture surfaces can be treated with a cleaning agent that is safe for wood and the finish. Again, do the cleaning outside whenever possible; if you have to work inside, *set up containment-like conditions with plenty of ventilation.* Before you do anything to a valuable antique, check with a dealer or restorer.

If you have upholstered couches or chairs that have a musty odor, you may want to get rid of them, whether you can see mold or not. This is particularly true if a piece of upholstered furniture has been wet for twenty-four to forty-eight hours or more, because thick cushions cannot be dried fast enough to prevent the growth of mold and other microbes. If any dry upholstered furniture is suspect, I recommend sampling the dust from the cushions for mold before the piece is thrown away or used again. If a piece of valuable furniture is contaminated, check with a furniture restorer to find out whether you can remove the fabric covering and stuffing and have the piece reupholstered.

Rugs and Carpeting
Area Rugs

A small area rug that has been flooded with clean (category 1) water should be removed from the house and allowed to dry in the sun (hang it up, don't put it on the ground!). Consider professional cleaning, particularly if the weather is humid or cloudy. If the rug and pad beneath have remained wet in place for twenty-four to forty-eight hours, it's likely that microbial growth has occurred, and you may want to replace both rug and pad. If the rug is valuable and

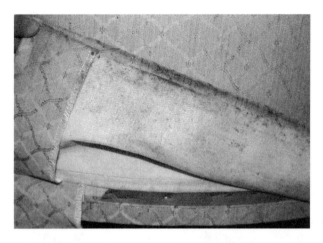

Hidden mold on the back of a chair cover. *Aspergillus* mold colonized the bottom edge of a chair cover that was close to a sliding glass door in a cold climate. The homeowner, who had severe respiratory problems, spent many hours in the chair, which was beside a heating system floor register that blew air across the moldy fabric. This was the only part of the chair that was contaminated with mold, and the cover could be removed and cleaned. The chair also had to be moved away from the cold glass door.

throwing it away is not an option, either have it professionally cleaned immediately or take it outside and use a wet vacuum to remove water as quickly as possible; then hang the rug in the sun. (I always think it's preferable to replace a rug contaminated with mold growth rather than try to have it cleaned.)

Whether you plan to discard a *dry* area rug that is contaminated or have it cleaned, it should be sealed in plastic before being carried out of the house. (Do *not* put a damp or wet rug in plastic if you intend to save the item.) Whoever is doing the work should wear a NIOSH N95 mask, and a window box fan should be running on exhaust before the work begins. It may also make sense to set up containment-like conditions if additional work will be necessary. Be sure to check the floor beneath, which should be HEPA-vacuumed and then probably washed with an appropriate cleaning agent, rinsed, and left to dry. Finally, be sure to get all the dust off nearby surfaces, including components of heating or cooling systems (such as the insides of duct boots, registers and grilles, and the tops and bottoms of baseboard convectors and the fin-tubes within).

Large area rugs that have been soaked by category 1 water or that have a musty odor should be evaluated by professionals (see chapter 11).

Wall-to-Wall Carpeting

If you suspect that a wall-to-wall carpet is moldy, you may be tempted to try to wash it yourself. I do not recommend doing so, for a variety of reasons, not the least of which is that if the carpet remains wet too long, mold and other microorganisms will grow in the carpet dust. Trying to clean a moldy carpet is a little like trying to clean a stinky sponge. You have to rinse the sponge out several times, and unless you use a powerful cleaner, the sponge will probably still smell after it dries. Unfortunately, a carpet and pad cannot be soaked and then wrung out, and the likelihood of removing all the mold by washing the carpet is minimal.

I think only certified professionals should clean wall-to-wall carpeting. (The Institute of Inspection, Cleaning, and Restoration Certification certifies carpet cleaners; see the Resource Guide.) Some carpet cleaners recommend applying fungicide or other antimicrobial solutions. I'm cautious about using such chemicals indoors, particularly on something like a carpet or rug, because residual amounts of whatever has been applied will be left in the dust and aerosolized when people walk across the surface (and some allergens may remain in the carpet even after the mold dies). If you do choose to have a chemical applied, be certain *first* that it has been approved for use on carpets (check with the chemical manufacturer). But in my opinion, if a carpet has to be treated with a biocide, it's probably too contaminated to be saved.

If you do have carpeting washed, make sure that the carpet dries out within twenty-four hours. If the weather is warm and dry, you can use a window exhaust fan and a floor fan to dry out the carpet (or use a window exhaust fan and rent a powerful blower called an *air mover*). If the weather is humid, you must use powerful dehumidifiers and air movers. If the weather is too cold to open windows, you will have to use the heat as well as dehumidify. Keep in mind that carpet laid on concrete will take longer to dry than carpet laid on wood. In the end, if the carpet has been washed and dried and still smells musty, it probably makes sense to replace it.

Hallways and Staircases

Don't forget that carpeting on stairs and in hallways can also be contaminated, and that every time people walk across these surfaces, allergens are released into the air. If rugs or carpets anywhere in your home contain mold growth, there's a good chance that the stairway and hallway rug and carpet will have mold as well, so these too should be removed and replaced. In several homes I investigated for families with mold allergies, all the moldy carpet was removed except for the carpeting on the stairs. The sensitized individuals continued to experience symptoms until that last section of carpeting was removed.

Removing Contaminated Carpeting

You may choose to have contaminated wall-to-wall carpeting and pad removed and replaced, and this is what I recommend. Again, removal should be undertaken by professionals, to minimize the spread of contaminants. All surfaces and heating components in the room, including the floor beneath the carpet and pad, should be cleaned of all dust with a HEPA vacuum before another pad and carpet are installed. If unfinished wood subflooring or plywood lies beneath the carpet, the surface might have to be sealed with varnish or paint. (Wood flooring must be allowed to dry before and after it is varnished or painted.) If the floor is concrete and it was soaked with water, a professional should determine when the concrete is dry enough for carpet installation.*

---

*See *Technical Bulletin: Adhesive Installation over Concrete Sub-floors* (Dalton, Ga.: Carpet and Rug Institute, 1999). Available at www.carpet-rug.com.

Vacuuming and Steam-Cleaning

HEPA-Vacuuming

One man called me because he was unable to live in his house. It was the first home he had owned, and it had wall-to-wall carpeting installed on a concrete slab. Shortly after moving in he began to have trouble breathing and sleeping at night. He was so fatigued during the day that he found it hard to concentrate at work. After about a year and a half he felt forced to move in with friends. He then paid a mold abatement (remediation) company to remove the carpeting and clean up the house, but he still couldn't move back in, because whenever he entered, his face felt itchy and the air felt "heavy." Even though the carpeting had been removed and the house supposedly cleaned, I found large numbers of residual mold spores in the dust left on the floor, as well as mold growing in the dust on the bottoms of the baseboard convectors. I recommended that all surfaces in the home, including the baseboard convectors, floors, and walls, be HEPA-vacuumed and damp-wiped (and the concrete be painted or coated with a sealant to adhere residual dust).

In fact, I recommend that all families with allergies and asthma use a HEPA vacuum for routine cleaning. Before purchasing a HEPA vacuum, check the ratings in *Consumer Reports*. Central vacuum systems can also be used for routine cleaning (but not mold cleanup), if they exhaust to the *exterior* of the home. Whether you use a HEPA or central vacuum, move the hose slowly and methodically back and forth, rather than jerk it around unevenly, to avoid kicking up dust. (One study compared four regular vacuum cleaners and two HEPA vacuum cleaners, and found that the levels of cat dander in the air after vacuuming were not very different, no doubt owing to the way in which people agitated the carpet while vacuuming.)

Steam Vapor Treatment

Consider having any upholstered furniture, carpeting, or rugs that have a slightly musty smell (but no mold growth) treated with steam vapor (SV). An SV cleaner looks like a vacuum cleaner, with canister, hose, wand, and head, but that's where the resemblance stops, because an SV machine does not vacuum up anything. The canister is really a kettle in which water is boiled and turned into vapor under pressure. Some vapor may condense in the hose, wand, and head, but what exits the cleaning head is very hot water vapor (hotter than 212°F, or 100°C), not liquid.

A traditional so-called steam carpet cleaner (the type that most professionals use and homeowners rent) is quite different. What comes out of the head is mostly *hot water* rather than steam. The water is squirted into the carpet and simultaneously extracted by a vacuum-cleaning suction device that is part of the cleaning head. If a carpet is "steam-cleaned" improperly by this method, the carpet and possibly the pad below become soaked with water. After a carpet has been properly treated with SV, however, it is barely damp, so there is less chance for fungal growth.

Because of the heat energy stored in pure hot vapor, a carpet properly treated with SV reaches a much higher temperature than one treated with hot water. Steam vapor treatment can kill up to 100 percent of insects (mites, fleas, silverfish, booklice, and spiders) living in the carpet. If done slowly enough, SV can also kill some of the mold spores and bacteria, and it can denature (destroy) many of the allergenic enzymes in spores, as well as protein allergens in mite body parts and cat and dog dander. From a health perspective (though not necessarily from the carpet's perspective), I think SV is one of the safer carpet treatments available, because there are no chemicals (pesticides, fungicides, or antimicrobials) involved.

A few extra cautions are worth noting. Rugs on hardwood floors should *not* be treated in place with SV, because the high temperature and moisture may damage the wood finish. Hang a rug to treat it, or support it off the floor. In the winter you may have to ventilate the space while using steam vapor, to avoid condensation problems; in the summer you should ventilate the space well, or use air conditioning or a dehumidifier to lower relative humidity levels. Remember that the temperature of the vapor is at least 212°F, so *be sure to check with the rug or carpet manufacturer before using SV* (the high temperature of the steam can affect fibers and dyes, particularly in some synthetic carpets and rugs). If you can't obtain information about the carpet itself, you should test a small area that is not normally visible, where any damage caused by the steam will not be that noticeable. Finally, follow the SV equipment manufacturer's directions carefully for safe use.

TIPS

Gather information about your options

- Before you tackle any mold problems, consult a number of resources as well as your physician.
- See the EPA's "A Brief Guide to Mold, Moisture, and Your Home" (listed in the Resource Guide).

Take precautions when tackling the job yourself

- Do not touch mold, or work with bleach or bleach solution or hazardous cleaning liquids, with bare hands; wear long protective gloves.
- When dealing with mold, wear protective gear, including plastic safety goggles and a properly fitted mask with a NIOSH rating of at least N95 (and remember, an N95 mask protects only against particulates, not against hazardous vapors). Depending on the size of the job, a higher level of personal protection may be required (see chapter 11).
- Protect your home and your health by setting up containment-like conditions to minimize the spread of moldy dust.
- When working with dry, moldy dust, use a HEPA vacuum rather than a regular vacuum cleaner (including wet/dry vac), which can emit allergens in the exhaust.

Know when to get help with the job

- If you have allergies or sensitivities to mold, the search for mold and even smaller cleanup jobs are best left to others who are not afflicted.

- I don't generally recommend that you hire a neighborhood teenager to help you clean the moldy basement or crawl space of your home. Hire an expert, insist on containment and the use of protective gear, and stay out of the way; your health may depend on it.

Get to the source of the problem

- Remember that mold is a living organism. Unless you minimize the conditions that led to its growth, the mold will likely reappear.
- Ask the important questions: Why is mold growing on items I own or on the surfaces in my home? What can I do to prevent fungal growth? When you have the answers, you will have made major progress in your battle against mold.

Chapter 11

PROFESSIONAL REMEDIATION

In an ideal world, if you were suffering from mold sensitivities, you would go to your physician with a health complaint, and the physician would try to determine whether mold was the cause of your symptoms. Then either you or an investigator whom you hired would find out if there was fungal growth in your home, and a professional remediation (abatement) company would clean up the problem—all of this work covered by your insurance company. After being eradicated, the mold would never grow back.

Unfortunately, few of these steps happen or are even feasible in the real world. It's rare that a physician can prove that a particular mold is the cause of health problems. Fungal growth is often concealed, and what mold *is* visible may not be the source of the spore exposure. And as we saw in chapter 4, many insurance companies refuse to pay for mold damage and remediation.

Identifying and solving mold problems can be difficult, and there aren't many people on your side. It is therefore vital to be as knowledgeable as you can when dealing with mold, whether it is a small spot of fungal growth on a bathroom ceiling or a large patch of mold growing on the basement drywall. In chapter 10 we discussed some of the ways in which homeowners who are not sensitized to mold can handle small mold cleanup jobs (level I). For levels II–V, professionals must be hired for the remediation.

Floods

Flooding, one of the biggest catastrophes that can happen to a homeowner, can lead to major mold problems.

Flooding with Category 1 Water

If you've had a major flood in your below-grade space or in a habitable room due to basement water or a plumbing disaster (category 1 water, defined at the beginning of chapter 10), and the walls, floor, floor covering (carpeting, rugs), and furniture have been soaked, call your insurance company immediately and hire professionals to deal with the cleanup.

I think any porous cushioned materials (including mattresses), disposable cardboard boxes, and other nonessential paper items that were saturated, as well as carpeting or rugs that have remained damp for twenty-four to forty-eight hours or more, should be discarded. This includes upholstered furniture, especially if the pieces smell. In other words, do not assume that wet items such as these can be dried out and will then be "good as new." (As noted previously, you may be able to have valuable pieces of furniture reupholstered; consult a furniture restoration specialist.)

If there has been half an inch of category 1 water in an above-grade or below-grade room, carpeting will get wet, and it should be dried out by professionals as quickly as possible (within twenty-four to forty-eight hours). If there are a few inches of water, it may enter the wall cavities to the same height as the water on the floor (it's a physical property of water to seek its own level). In either case, water will flow by capillary action up the walls and dampen plaster, drywall, or paneling, as well as insulation, above the water level. Unless you live in an arid climate, this kind of situation can set the stage for hidden proliferation of mold within wall cavities and in floor dust in or under the carpet and pad. (It will *rarely* be adequate to dry only the exposed room surfaces and hope that the wall cavities, if wet, will dry before mold grows within.)

The demolition and removal of saturated or very moldy walls and ceilings should be handled by trained mold remediators. Walls filled with fiberglass or cellulose insulation that become soaked by plumbing leaks or floor water should be opened up where the leakage occurred or to the flood line, plus an additional twelve inches or more above. All wet wall insulation should be removed, and the remaining wall materials should be allowed to dry (check with a moisture meter) before new insulation is installed and the wall framing is closed up. (Wood framing, when dry, may have to be coated with sealant.) Remember, I recommend using sheet-foam insulation, and not fiberglass or cellulose insulation, in below-grade walls.

If you aren't sure whether certain wall cavities were affected by a floor flood or plumbing leak, some openings should be carefully cut in the plaster or drywall near the floor or under the leak. It is a lot simpler to patch a few holes than to remove drywall and paneling weeks or months later, after they start to stink. Even after minor floor flooding, mold might be found behind wood or plastic baseboards fastened to drywall, because floor water can soak into the paper by capillary action and be trapped by the baseboard. (In some new below-grade construction I have found *Stachybotrys* mold growing completely hidden behind the baseboard.) If you or anyone in your family has mold sensitivities, leave this exploratory work to professionals.

If any components of the hot-air furnace or air-conditioning system have been under water, the system should be immediately evaluated by professionals. At a minimum, the interior of the unit should be cleaned and sanitized with an EPA-approved product, and any fiberglass lining material should be removed and replaced professionally. The filter should also be replaced.

Finally, if the flood occurred in an above-grade room, check the basement or crawl space beneath, in case mold problems are developing there as well, because just one major oversight like failing to check this area can lead to trouble.

A heat pump return that was submerged. This is the return box under a heat pump fan coil in a basement that flooded several times. The sheet-metal box rests directly on the floor, and there are two very clear horizontal water lines on the metal. One line is at about the middle of the box, and the other is about four inches below. On the floor to the left of the box is the condensate pump. Dirt on the dark wall of the pump was also from floor water. Unfortunately, no one ever bothered to clean out the interior of the return box, where the fiberglass liner was filthy and completely full of mold years after the flooding. Over that time period at least two consecutive owners became highly sensitized to mold, I suspect from breathing the air that came from the system. Neither was able to continue living in the home.

I am familiar with one home in which most of the inside was renovated about two years after major flooding occurred, but the original subfloor and maple floor, which had been saturated, were left in place. The developer never realized that mold had grown between the two wood floor layers, and mold-eating mites had moved in. Eventually the mold died, but for twelve years, as the new owners walked across the carpet and compressed the warped maple flooring, mold and mite allergens were emitted from the gaps between the subfloor boards in the basement ceiling below. One of the new owners was highly allergic to the dust and developed asthma after spending many weekend hours in his basement shop.

After flooding with category 1 water, it's important to use the right equipment to minimize microbial growth. In most cases, if the weather is humid, the use of fans (or professional air movers) alone will be inadequate; powerful dehumidifiers, which remove moisture from the air, *must* be used to dry the area. Some professional remediators keep the relative humidity as low as 10 percent with dehumidifiers, in order to dry a flooded space rapidly. Other remediators use fans or air movers that haven't been cleaned since their last use, which can spread mold from the previous job into your indoor environment, so insist that your contractor use clean equipment. In fact, if the basement is full of moldy goods and surfaces, fans should not be used at all until the basement has been isolated (contained) from the rest of the house and moldy items have been removed, because the increased airflows will kick up moldy dust and spread the spores upstairs.

A wet/dry vac can get rid of water quickly. Typical homeowner wet vacs should not be used for gray or black water, however, because they can create aerosols of microorganisms that if inhaled may lead to illness. After the floodwater is gone, dry, moldy dust should be vacuumed using a HEPA (high-efficiency particulate arrestance) device. Regular vacuum cleaners, including wet/dry vacs, should *never* be used *indoors* when working with dry, contaminated dust, because allergens can be emitted in the exhaust air. (After any flooding, it is a good idea as a precaution to HEPA-vacuum *all* surfaces in the entire house, but particularly in areas where there has been worker traffic.)

Flooding or Leakage with Gray or Black Water

After major leakage or flooding with gray or black water, remediation *must* be done by trained professionals. If a sewer or septic system has backed up, call a professional fire and water damage restoration company to handle the cleanup. Carpeting, the pad beneath, and furniture with cushions that have been contaminated in any way with black water should be thrown out and replaced. If a furnace, AC, or indoor heat pump unit has been submerged in black water, it may also have to be replaced. Immediately after category 3 flooding, contact a heating or AC contractor and your local health department for advice, and do not operate any equipment until it has been deemed by a professional to be safe for use. (Note: A furnace heats air, whereas a boiler heats water. If a boiler has been submerged in any category of water, it should be professionally cleaned and repaired as needed, but it may not be necessary to replace it. Again, contact a heating technician before operating the unit.)

Professional Remediators

How do you know when a mold remediator is qualified and insured, and charges reasonable rates? When choosing a contractor it can be helpful to gather information about your options from a variety of sources, including recommendations from past clients, references from your insurance company, and advice in professional guides (for a listing of useful publications, see the Resource Guide at the end of the book). But if the mold was caused by a flood, time is of the essence, so don't delay in making decisions. And once you have chosen a contractor, verify that the company has liability insurance for mold remediation.

In many areas mitigators who are licensed to remove asbestos and lead paint have gone into mold remediation work as well. The methodologies are comparable, but specialized training in mold removal methods is important in qualifying someone for this kind of work. Some of the organizations that provide training and certification for mold remediators are listed in the Resource Guide at the end of the book. In addition, in the near future, a number of organizations will be publishing guidelines for mold remediation.

Containment Practices

If there is a major mold problem in one part of a house, proper containment measures *must* be used to prevent the spread of contami-

Stachybotrys mold on the backside of drywall in a basement that flooded. A finished basement flooded with water, and fiberglass insulation remained damp for weeks before it was removed and the extensive mold problem was discovered. *Stachybotrys* mold grew on the backside of the drywall, but on the front (painted) side there was no indication of any fungal growth. Note the stains on the vertical stud at the center, indicating capillary flow of water.

nants. It's important, therefore, to ask the remediator specific questions about containment and cleanup practices.

Professional remediators should isolate the area using plastic sheeting (treated with flame retardant) and reduce the air pressure in the isolated area by using powerful HEPA-filtered exhaust blowers called *negative air machines*. Remediators may also use air scrubbers (high-volume HEPA air filters) to reduce the concentrations of mold contaminants within the work area. Remediators usually set up a separate area where workers change into their remediation gear: safety goggles, protective suits, gloves, and booties, as well as respirators. Since very moldy dust may be generated, wherever possible workers should enter and exit the containment area directly through an exterior door, to avoid passing through other areas in the home. Floors, walls, heating or cooling grilles, or any other items within the work area may have to be covered. Smoke testing may be used to ascertain whether air is flowing from the containment area into other rooms.

As an additional precaution, you may wish to hang plastic sheets over doorways (see chapter 10) between the containment area and other rooms in the house. It also makes sense to remove area rugs, particularly where workers will be walking back and forth, and to tape nonslip protective coverings over these paths.

In order to provide proper containment, air must be continuously blown out of the contained work area (and provisions made for air from the rest of the building to infiltrate the work area). Unless the exhausted air is HEPA-filtered, mold spores may move with wind into indoor environments close by.

Keep in mind that most moldy construction materials are considered rubbish, and there are no requirements that I know of for secured disposal, as there are for lead paint and asbestos. Of course, everything that gets thrown out (including your everyday garbage), if it goes to a landfill, is eventually attacked by microorganisms that include mold. But that doesn't mean you shouldn't be cautious in the interest of protecting your neighbors. Tossing moldy materials into an unprotected dumpster can aerosolize contaminants. Moldy materials should therefore be sealed in heavy-duty plastic bags before removal from the containment area and disposal.

Other People Who May Be Involved

Other people can unwillingly become involved in a number of ways when you are dealing with leaks, floods, and mold. When there are moisture or mold problems, *immediately* notify other people who may be responsible or affected, including your neighbors.

If you live in an apartment or townhouse that shares a wall with another unit and you experience plumbing or roof leaks that lead to mold growth in your home, the source of the leak may turn out to be on someone else's property. If the leaking pipe is located in the bathroom floor of the unit above yours, the occupant of that unit, and his or her insurance company, may be responsible for covering the cost of the damage to your property. A neighbor might claim that the mold in his or her basement is due to a leak from a pipe in your adjoining wall, although conditions of high relative humidity are the major culprit in such cases. If you have a leak in your basement, on the other hand, and the water invades your neighbor's basement as well, then you are responsible for the damage, so be sure to check adjacent spaces as soon as possible after a leak has occurred.

If you live in a condominium and the roof leaks, the condominium association is usually responsible. If it looks as if the association will not respond immediately, you may want to take action to minimize the likelihood that mold will grow in your unit. Consider cutting a hole in your ceiling and putting a bucket underneath, so roof water (category 1 water) does not soak into the ceiling or accumulate in your carpet. Don't wait for the drywall to sag and the ceiling to collapse or for mold to grow before taking commonsense steps to defend your interests and health. (On the other hand, if there are signs that leaks have occurred previously in the same location, you must be *very careful* about exposing established mold growth!)

Your Insurance Company

Check the current version of your insurance policy to see under what circumstances mold testing and cleanup are included or excluded (see chapter 4). Many insurance policies will pay for damage associated with covered losses, such as those due to plumbing leaks, but will not cover losses due to mold growth caused by improper or sloppy maintenance practices. In some cases mold insurance is available for an extra premium.

Always be thorough in your investigation of the extent of the spread of the water, as well as the direct relationship between the water event and the mold growth. *Immediately* notify your insurance agent of any moisture or mold problems. Take photographs to document water problems or mold growth. Remember that an area affected by leaks or flooding should be dried out as quickly as possible to minimize mold growth, so depend on your documentation rather than leave things wet to show the insurance adjuster.

TIPS

Floods can lead to mold problems

- Call professionals to clean wall-to-wall carpeting that has been flooded.
- Walls that were affected by flooding may have to be dried out or even replaced; wet insulation should be discarded.
- Any hot-air furnace or air-conditioning system that was under clean water (category 1) should be professionally cleaned and sanitized; if under black water (category 3), it should probably be replaced. Consult your health department and a heating contractor for advice.
- Professionals should probably handle the cleanup of major floods, even those involving clean water; professionals *must* handle the cleanup of major leakage or flooding with gray or black water.
- Carpeting, rugs, and pads that have been contaminated with black water must be discarded.*

Be selective when hiring a professional remediator

- Gather information about your options before hiring a remediation company.
- Insist on proper containment to protect yourself, your home, your family, and your neighbors.

Others may be involved

- If you are a tenant or a condominium owner, keep careful records of your experiences with water intrusion, mold, or a musty odor.

Standard and Reference Guide for Professional Water Damage Restoration S500-94 (Vancouver, Wash.: Institute of Inspection, Cleaning, and Restoration Certification, 1995).

- If you are a property manager or a landlord, properly maintain buildings and systems to minimize the chances of mold growth, and deal with water or mold problems quickly, safely, correctly, and sympathetically. Keep accurate records of complaints, your responses, and any repairs made.

Your insurance company may cover some of the expenses

- Check your current insurance policy to see if damage from mold growth and the cost of mold remediation work are covered.
- Contact your insurance company immediately after a flood, and document any water damage or mold growth.
- Keep in frequent touch with your insurance company to be sure that a mold or water problem is dealt with quickly, efficiently, and professionally.

CONCLUSION

Some people are ready to flee their homes the moment they see a fuzzy black patch on a wall, while others are oblivious to walls that are literally blanketed with mold. People who are not concerned about mold think those who worry about it are overreacting. People who are not affected by mold think those who complain of symptoms due to mold exposures are hypochondriacs or malingerers. And the afflicted feel frustrated because they think other people discount their suffering.

Most environmental threats to health have had a controversial history, with believers and prophets of doom on one side and naysayers on the other. For example, thousands of years ago some people understood that lead could be hazardous, but only recently has the medical community accepted that lead can seriously impair children's development. A similar story applies to asbestos, a naturally occurring mineral that is now known to be a carcinogen. And to this day, despite numerous rigorous studies, many people deny that smoking is a major cause of lung cancer and heart disease.

Now a battle about mold is being waged in the media and in the scientific community. Hundreds of scientific studies of indoor environments conducted all over the world have shown a relationship between experiencing health symptoms and occupying damp buildings (where mold, bacteria, and other microorganisms can flourish). At the same time, some experts continue to claim there is no scientific evidence to prove a link. In my opinion such statements of denial are based on a narrow interpretation of what is relevant. While it is true that no one has proven definitively that exposure to mycotoxins from *spore inhalation* has caused specific illnesses, time and time again investigators have found that wherever there are chronically elevated concentrations of airborne mold spores or other bioaerosols indoors, a certain percentage of the people who spend time there (most often women) suffer allergy, asthma, and respiratory and other chronic symptoms. When the mold (along with other contaminants) is eliminated, people feel better. How much more proof do we need?

My experience has taught me that people report health effects even in the absence of *elevated* concentrations of mold spores indoors. In the hundreds of moldy buildings where I've sampled air and dust, sometimes the levels of spores were outrageously high, and other times the levels were not that different from levels I found outdoors. Nonetheless, in nearly every "sick building" people's health improved after the mold and other microorganisms were eliminated from their indoor environments. This is why I firmly believe that the most important measure of the impact of mold on indoor air quality is the well-being (or lack of well-being) of the people spending time in the space, and not only the concentration or type of contaminants present.

You should now have some answers, rather than more uncertainty, and be able to move past the media hype to recognize that yes, indoor mold is a problem that can affect health, but you don't have to be powerless or helpless. I hope you can now take some of the steps necessary to improve the indoor air quality of your home and thus improve your own health and the health of those you love.

RESOURCE GUIDE

A listing in this Resource Guide does not constitute an endorsement by the authors or the publisher; the Resource Guide is intended to help readers gather information and make up their own minds as to which organizations, products, or services may help them deal with mold problems.

Organizations and Websites

- American Academy of Allergy, Asthma and Immunology (AAAAI) (www.aaaai.org).

- American Academy of Pediatrics (www.aap.org).

- American Industrial Hygiene Association (AIHA) (703-849-8888; www.aiha.org).

- American Lung Association (1-800-LUNG-USA [586-4872]; www.lungusa.org).

- American Society of Home Inspectors (ASHI) (Des Plaines, Ill.; 800-743-2744; www.ashi.com). Has chapters nationwide and will provide the names of member home inspectors.

- Association of Energy Engineers (AEE) (Atlanta, Ga.; 770-447-5083; www.aeecenter.org). Certifies indoor air quality professionals.

- Association of Specialists in Cleaning and Restoration (ASCR) (Millersville, Md.; 800-272-7012; www.ascr.org).

- Asthma and Allergy Foundation of America (AAFA) (Washington, D.C.; 1-800-7-ASTHMA; www.aafa.org). Has chapters in a number of states and offers lectures and educational programs for people who have asthma.

- California Department of Health Services (www.dhs.cahwnet.gov/default.htm).

- Canada Mortgage and Housing Corporation (CMHC) (613-748-2367; www.cmhc-schl.gc.ca).

- Centers for Disease Control and Prevention (CDC), National Center for Environmental Health (www.cdc.gov/).

- Children's Environmental Health Network (www.cehn.org/New.html).

- Indoor Air Quality Association (IAQA) (12339 Carroll Avenue, Rockville, Md. 20852; 301-231-8388; www.iaqa.org). Offers courses for remediators and investigators, and lists Internet resources on the website.

- Indoor Air Quality Information Clearing House (800-438-4318). The EPA's indoor air quality information hotline.

- Institute of Inspection, Cleaning, and Restoration Certification (IICRC) (360-693-5675; www.iicrc.org). Certifies carpet cleaners and publishes "IICRC Standard for Professional Mold Remediation S520."

- National Air Duct Cleaners Association (NADCA) (Washington, D.C.; 202-737-2926; www.nadca.com). Disseminates information, sets standards, and encourages ethical practices in the duct-cleaning industry.

- New York City Department of Health, Bureau of Environmental and Occupational Disease Epidemiology (www.ci.nyc.ny.us/html/doh/html/epi/moldrpt1.html).

- U.S. Environmental Protection Agency (EPA), National Center for Environmental Publications (NSCEP) (P.O. Box 42419, Cincinnati, Ohio 42419; 800-490-9198; www.epa.gov/iaq). Publishes materials about mold and indoor air quality.

- www.MoldUpdate.com. Offers information about mold.

Products and Services

Supplies and Equipment

- AllergyZone LLC (9812 Shelbyville Road, Suite 2, Louisville, Ky. 40223; 800-704-2111; www.allergyzone.com). Manufactures a disposable, high-efficiency furnace filter with a MERV rating of 12.

- ARAMSCO, Inc. (1655 Imperial Way, P.O. Box 29, Thorofare, N.J. 08086-0029; 800-767-6933; www.aramsco.com). Sells Dust Shield™ adjustable poles (for containment enclosures) and sets of poles (for a three-sided booth or Zip Wall™).

- Denny Sales Corporation (3500 Gateway Drive, Pompano Beach, Fla. 33069; 800-327-6616). For Dennyfoil, a paper–aluminum foil laminate that can be used to temporarily cover surfaces.

- Headrick Building Products, Inc. (5775 Bethelview Road, Cumming, Ga. 30040; 678-513-7242; www.headrick.net). Manufactures ridge and soffit vents.

- Home Environmental (Lexington, Mass.; 781-862-2873). For steam vapor machines, HEPA vacuums, and other allergy products.

- Moisture Solutions LLC (5900 Monona Drive, Suite 400, Madison, Wisc. 53716; 608-663-4735; www.moisture-solutions.com). Sells equipment that uses low-wattage electrical energy to minimize rising damp in foundations.

- Professional Equipment (Hauppauge, N.Y.; 800-334-9291). For testing and other inspection equipment, including moisture meters.

- Statewide Supply (Cross Plains, Wisc.; 800-553-5573). For a floor water alarm by Gizmode Innovations.

- Therma-Stor Products (Madison, Wisc.; 800-533-7533). For dehumidifiers.

Laboratories That Will Analyze Dust Samples from Carpets and Other Surfaces

- Aerotech Laboratory, Inc. (Phoenix, Ariz.; 800-651-4802; www.aerotechlabs.com).

- American Industrial Hygiene Association (www.aiha.org). Offers a list of labs accredited by EMPAT (Environmental Microbiology Proficiency Analytical Testing).

- DACI Laboratory, Johns Hopkins University Asthma and Allergy Center (Baltimore, Md.; 800-344-3224).

- Northeast Laboratory (Waterville, Maine; 800-244-8378; www.NeLabServices.com).

Mold Consultation and Remediation

- Environmental Testing and Technology, Inc. (Carlsbad, Calif.; 760-804-9400). Indoor air quality testing.

- Envirotech (Stoneham, Mass.; 781-279-2900). Specializes in duct cleaning and mold remediation.

- May Indoor Air Investigations LLC (Cambridge, Mass.; 617-354-1055; www.MayIndoorAir.com). Indoor air quality, moisture, mold, and odor investigations.

- www.MyHouseIsKillingMe.com. Offers a geographical list of indoor air quality consultants.

Institutions That Train Mold Consultants and Remediators

- ACGIH Professional Learning Center (Cincinnati, Ohio; 513-742-2020; hazard.com/mail/safety-new/msg00050.html).

- Indoor Air Quality Association (IAQA) (12339 Carroll Avenue, Rockville, Md. 20852; 301-231-8388; www.iaqa.org).

- MidAtlantic Environmental Hygiene Center (MEHRC) (3624 Market Street, First Floor East, Philadelphia, Pa. 19104; 215-387-4096; www.mehrc.com).

- Restoration Consultants (3463 Ramona Avenue, Suite 18, Sacramento, Calif. 95026; 888-617-3266; www.restcon.com).

Other Services

- McClaughlin Upholstering Company, Inc. (Everett, Mass.; 617-389-0761). Specializes in reupholstering and restoring antiques.

- Munters Corporation (www.muntersamerica.com). Has offices nationwide and specializes in emergency flood remediation (dehumidification, drying).

- Neutocrete Systems, Inc. (888-799-9997). Installs a masonry coating over dirt floors in crawl spaces.

- Northeast Document Conservation Center (100 Brickstone Square, Andover, Mass. 01810-1494; 978-470-1010; www.nedcc.org). Gives advice on document conservation.

Publications

- "Biological Pollutants in Your Home," EPA 402-F-90-102, January 1990 (www.epa.gov/iaq, or IAQ INFO, P.O. Box 37133, Washington, D.C. 20013-7133; 800-438-4318).

- "A Brief Guide to Mold, Moisture, and Your Home," EPA 402-K-02-003, Summer 2002 (www.epa.gov/iaq, or IAQ INFO, P.O. Box 37133, Washington, D.C. 20013-7133; 800-438-4318).

- "Building Air Quality: A Guideline for Building Owners and Facility Managers," U.S. Environmental Protection Agency and the National Institute for Occupational Safety and Health, 1991.

- "Clean-up Procedures for Mold in Houses," Canada Mortgage and Housing Corporation (613-748-2367; www.cmhc-schl.gc.ca).

- "Cleaning up Your House after a Flood," Canada Mortgage and Housing Corporation (613-748-2367; www.cmhc-schl.gc.ca).

- "Flood Cleanup: Avoiding Indoor Air Quality Problems," EPA 402-F-93-005, August 1993 (www.epa.gov/iaq, or IAQ INFO, P.O. Box 37133, Washington, D.C. 20013-7133; 800-438-4318).

- "Fungi and Indoor Air Quality," by Sandra V. McNeel, D.V.M., and Richard A. Kreustzer, M.D., *Health and Environment Digest* (California Department of Health Services, Environmental Health Investigations Branch), vol. 10, no. 2 (May/June 1996):9–12.

- *Guidelines on Assessment and Remediation of Fungi in Indoor Environments* (New York: New York City Department of Health and Mental Hygiene, Bureau of Environmental and Occupational Disease Epidemiology, 2000).

- "Handbook of Pediatric Environmental Health," American Academy of Pediatrics, 1999 (www.aap.org).

- "IICRC S001 Carpet Cleaning Standard," Institute of Inspection, Cleaning, and Restoration Certification, Vancouver, Wash., 1997.

- "Molds, Toxic Molds, and Indoor Air Quality," by Pamela J. Davis, *CRB Note*, vol. 8, no. 1 (March 2001), ISBN 1-58703-133-7.

- *My House Is Killing Me! The Home Guide for Families with Allergies and Asthma*, by Jeffrey C. May (Baltimore: Johns Hopkins University Press, 2001).

- *My Office Is Killing Me!* by Jeffrey C. May (Baltimore: Johns Hopkins University Press, forthcoming).

- "Policy Statement: Toxic Effects of Indoor Molds," *Pediatrics* (American Academy of Pediatrics), vol. 101, no. 4 (April 1998):712–14 (www.aap.org/policy/re9736.html).

- "Should You Have the Air Ducts in Your Home Cleaned?" EPA 402-K-97-002, October 1997 (www.epa.gov/iaq, or IAQ INFO, P.O. Box 37133, Washington, D.C. 20013-7133; 800-438-4318).

- "Use and Care of Home Humidifiers—Indoor Air Facts No. 8," EPA 402-F-91-101, February 1991 (www.epa.gov/iaq, or IAQ INFO, P.O. Box 37133, Washington, D.C. 20013-7133; 800-438-4318).

INDEX

acquired immunity, 37, 40

actinomycetes, 85

aflatoxins, 47–48

air conditioning systems, 27–28, 35, 84, 86–90, 104, 161; cleaning of, 190, 197; cooling coils of, 89; filtration of, 95–98; leaks in, 92; window or wall units, 101–2, 104–5, 120–21. *See also* evaporative coolers

air conveyance systems, 93–94; mold spores in, 161. *See also* air conditioning systems; *and types of heating systems*

air exchange rate, 29, 32

airflows, 24–29

air movers, 182–83

air pressure, 26–27, 33–35, 91, 135, 140, 174

air sampling, 156, 157–59; results of, 160–65

alcohol, 178–80

allergens, 161, 176. *See also* mold testing

allergic reactions, 42

allergic rhinitis, 37, 39

allergies, 38–40, 55, 70–71, 129, 130

Alternaria, 10, 40, 88, 132, 160

American Academy of Allergy, Asthma, and Immunology (AAAAI), 161–62

American Industrial Hygiene Association (AIHA), 160

American Society of Home Inspectors (ASHI), 155, 171

anaphylaxis, 42

angioedema, 42

animals: dander, 80; pet urine, 110

area rugs, 180–82, 185

Armillaria mellea, 9

arrestance, 96

aspen, 8

aspergillosis, 51

Aspergillus, 10, 14, 23, 42, 160; in attic ducts, 88; on baseboard convectors, 99; in basement ceilings, 73; on doors, 179; on furniture, 106, 107, 181; illustrations of, 13, 107, 157, 179, 181; spores of, 13, 50

Aspergillus flavus, 43, 47

Aspergillus fumigatus, 40, 51

Aspergillus ochraceous, 48

asthma, 39–40, 55, 108, 163, 178, 192

atranones, 48

attics, 28–29, 132–37, 151; ductwork in, 134, 151; odors in, 136–37, 152; ventilation of, 133–34, 152

bacteria, 38, 46, 85

bacteriacide, 95

Ballard, Melinda, 53–54

barn boards, 122

baseboard convectors, 99, 124, 178, 182; cleaning of, 104

basement flooding, 18–19, 20, 189–90

basements: finished, 70–74, 82; unfinished, 65, 68–70, 78, 81

bathrooms, 127–28, 130–31; bathtubs, 112–13; exhaust for, 130, 134, 152;